Building Pathology and Rehabilitation

Volume 1

Series Editors
V. Peixoto de Freitas
Aníbal Costa
J.M.P.Q. Delgado

For further volumes:
http://www.springer.com/series/10019

V. Peixoto de Freitas · J.M.P.Q. Delgado
Editors

Hygrothermal Behavior, Building Pathology and Durability

Springer

Editors
V. Peixoto de Freitas
Building Physics Laboratory
Department of Civil Engineering
Faculdade de Engenharia
Universidade do Porto
Porto
Portugal

J.M.P.Q. Delgado
Building Physics Laboratory
Department of Civil Engineering
Faculdade de Engenharia
Universidade do Porto
Porto
Portugal

ISBN 978-3-642-43788-5 ISBN 978-3-642-31158-1 (eBook)
DOI 10.1007/978-3-642-31158-1
Springer Heidelberg New York Dordrecht London

Softcover reprint of the hardcover 1st edition 2013

Printed on acid-free paper

Springer is part of Springer Science+Business Media (www.springer.com)

Preface

In the distant past, construction solutions were validated empirically through several years of experience, whereas with the onset of further industrialization of the construction process, it was recognized within the construction community that a performance-based selection of materials, components, and systems was required if innovation was to be fostered and progress in the construction domain achieved. However, it was equally apparent to those promoting such novel approaches that the selection on the basis of understanding of performance requirements could only be met if the results of research and development were made available and indeed exploitable by practitioners.

Building pathologies originated by moisture are frequently responsible for the degradation of building materials or components and can affect users' health, comfort, and durability. The solutions for treating moisture-related pathologies are complex and, many times, of difficult implementation. Several of these pathologies are due to innovative techniques and materials combined with new details and predicted performance. The knowledge of the physical processes that define hygrothermal behavior allows for the prediction of a building response to climatic solicitation and for the selection of envelope solutions that will lead to required feasibility.

Rehabilitation is a strategic area that is concerned not only with historic buildings, but also with other buildings that have been in use for some time and need to be adapted to the demands of the present. The following areas should also be considered: rehabilitation and hygrothermal performance of buildings, diagnosis, measurements in-field and in laboratory, hygrothermal advanced simulation, energy efficiency, risk analysis, and technology.

The main purpose of this book, *Hygrothermal Behavior, Building Pathology and Durability*, is to provide a collection of recent research works to contribute to the systematization and dissemination of knowledge related to construction pathology, hygrothermal behaviour of buildings, numerical simulation and durability, and, simultaneously, to show the most recent advances in this domain. It includes a set of new developments in the field of building physics and hygrothermal behavior and building pathology versus durability. This book is divided into several chapters that intend to be a resume of the current state of knowledge for benefit of professional colleagues, scientists, students, practitioners, lecturers, and other interested parties to network.

V. Peixoto de Freitas
J.M.P.Q. Delgado

Contents

Reducing the Risk of Microbial Growth on Insulated Walls by Improving the Properties of the Surface Materials

Martin Krus, Cornelia Fitz and Klaus Sedlbauer

Abstract Staining of façades due to microbial growth has been risen in recent years. Especially walls with external insulation systems are affected because the low thermal mass of the exterior render combined with a high thermal resistance of the insulation layer leads to frequent overcooling of the render's surface by long-wave radiation exchange with the sky. Condensation forming on the overcooled façade is of major importance for microbial growth. Therefore, the best way to prevent growth is to reduce the frequency of condensation by limiting the periods of overcooling. This could be achieved by increasing the thermal inertia of the exterior layer by using massive rendering or the addition of phase change materials (PCM) or by applying low IR emissivity (low-E) paint coats. According to the present state of knowledge not the amount of condensation water is decisive for the risk of microbial growth but the condensation water on the external surface, which is available for micro-organisms in the initial phase of growth. Absorbent substrates could offer remarkable advantages in this respect. The performance of novel rendering systems provided with one of these measures or with a combination has been investigated by field tests as well as by hygrothermal simulations.

Keywords Microbial growth · Phase change materials · Low-E coating · Exterior condensation

M. Krus (✉) · C. Fitz · K. Sedlbauer
Holzkirchen Branch, Fraunhofer Institute for Building Physics (IBP), Valley, Germany
e-mail: Martin.Krus@ibp.fraunhofer.de

C. Fitz
e-mail: cornelia.fitz@ibp.fraunhofer.de

K. Sedlbauer
e-mail: klaus.sedlbauer@ibp.fraunhofer.de

V. P. de Freitas and J. M. P. Q. Delgado (eds.), *Hygrothermal Behavior, Building Pathology and Durability*, Building Pathology and Rehabilitation 1,
DOI: 10.1007/978-3-642-31158-1_1,

1 Introduction

In recent times more and more complaints about microbial growth on façades have been noticed (Blaich 1999). Figure 1 displays a façade with the typical microbial growth of fungi and algae. The surface above the tilted window clearly shows that even the ventilation behaviour of the residents can have an influence on the microbial growth. The effect of thermal bridges is also clearly visible. The slightly higher surface temperature causes an interruption of the massive fungal growth above the window. The surface below the window shows predominantly the growth of algae.

This growth occurs mostly in the first years after completion which leads to displeasure from the building owner. Figure 2 shows a photographic view of a building, on which the right side has been insulated about 1 year before taking this picture. The left side has been left unchanged. It is obvious that the higher insulation standard has caused the growth of algae or mould.

Since biocidal finishing of surfaces should be an extraordinary solution for environmental reasons and due to the temporary efficiency, new possibilities must be found to avoid the growth of algae or fungi preferably by physical measures. For this purpose, comprehensive analyses of façade surfaces are conducted on the outdoor testing site of the Fraunhofer Institute for Building Physics (IBP). Because there are different opinions about the higher sensitivity of walls with ETICS (external thermal insulation composite system) compared to monolithic walls and also the physical causes are not yet clarified, for a detailed analysis different wall constructions have been built at the test field. The whole variety of interesting variations, however, cannot be studied by field tests. Instead calculations can be carried out with the hygrothermal program WUFI® modified for this purpose. After validation it allows the rapid and cost-efficient assessment of numerous further variations to evaluate the influence of different factors, e.g. orientation, thickness of thermal insulation layers or enhancement of the thermal storage capacity by using PCM effects.

2 Growth Conditions

Algae and mould fungi are spread over the whole world and they are a major part of the ecosystem. Humidity is fundamental for algal growth as it is needed for photosynthesis. Because algae don't have any roots, the water uptake must occur directly through the cell wall by osmosis. The growth limit for green algae is at least 70–80 % RH (Denffer 1983) and for blue algae 100 % RH (liquid water) (Scherer 1993). Algae can survive dry periods without any harm and can restart their growth when enough humidity is available. Therefore, drying of façades during the day is not sufficient to prevent algal growth. Venzmer describes the optimal growth conditions on facades for green algae within a temperature range from 0–40 °C (Venzmer 2001). For the growth of mould depending on the temperature and the quality of the substrate a relative humidity on the surface of between 75–99 % is required (Sedlbauer 2001).

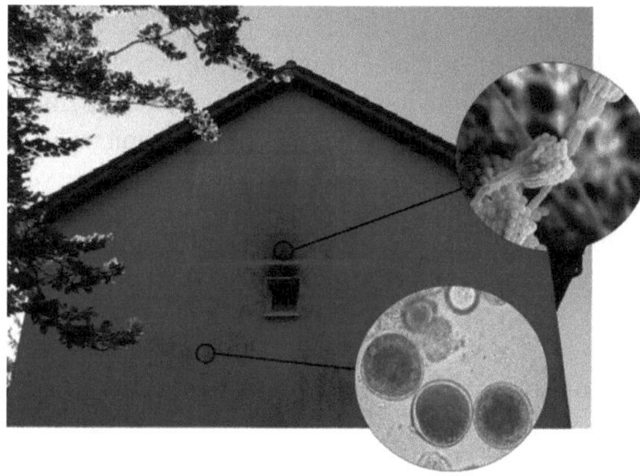

Fig. 1 Growth of algae on a façade with an external thermal insulation composite system (ETICS) and fungal growth above the window

Fig. 2 Photographic view of a building, on which the *right* side has been insulated about 1 year before taking this picture. It is obvious that the higher insulation standard has caused the growth of algae or mould

3 Assessment of the Risk for Microbial Growth

Besides the substrate effect on the surface, microbial growth on intensively insulated external building components exposed to weathering is a consequence of increased moisture loads on the surface and the resulting favorable conditions for growth (Krus and Sedlbauer 2003; Hofbauer et al. 2003; Denffer von 1983). As mentioned above suitable temperature and humidity conditions at the outer surface of walls are necessary for biological growth. Beneath wetting by driving rain, condensation occurs in consequence of long wave radiation during clear nights resulting in temperatures below the dew point of the air. The importance of this wetting mechanism is obvious with regard to the occurrence of microorganisms mostly on the northern sides of buildings where driving rain rarely occurs. Therefore, the periods of surface condensation and the accumulated degree of cooling below dew point temperature are taken as criterion to classify the results (Krus and Sedlbauer 2003).

The most essential criterion for the risk of microbial growth on façades is the availability of sufficient amounts of water. According to the present state of knowledge, it is obvious that the amount of condensation water is not decisive for the risk of microbial growth, but the condensation water on the external surface, which is available for micro-organisms in the initial phase of growth. For this reason, where possible, the amount of water on the surface was determined by dabbing the water with a tissue (Krus and Fitz 2007) and used for the assessment of growth risk, too.

4 Computational Investigations

Only the modelling of the transient processes allows the variation of all parameters of interest presented in this paper. Computations are carried out using the simulation program WUFI® (Künzel 1994), which was developed at the IBP and has been validated for numerous hygrothermal problems. To allow the comparison of the computed results (surface temperature and surface humidity) of individual wall constructions with measured data, calculations are carried out on the basis of the boundary conditions of the exterior climate recorded during the real weather exposure. Therefore, the following results refer to the climate in Holzkirchen. All calculations are done with an adapted version of WUFI® for the consideration of long wave irradiation (Künzel et al. 2002) with climate data for Holzkirchen in order to compare the surface temperatures with the dew point temperature of outdoor air.

The following two construction types are compared by hygrothermal calculations. The first one is a wall made of concrete with an ETICS. The other wall consists of aerated concrete. Both walls have a similar thermal transmittance of about 0.35 W/m^2K. Starting from a standard ETICS construction different surface properties were examined. Beside the colour (impacting the radiation coefficient) of the rendering the influence of the long wave emissivity was of special interest. In addition numerous field tests have been conducted on walls with ETICS facing to the north as well as to the west to verify the calculated results.

The most critical seasons for biological growth on external walls are autumn and spring. Winter and summer do not offer any optimal climate for the growth of algae and fungi (Hofbauer et al. 2006).

4.1 Validation

For a clear night the evaluation of the hygrothermal calculation software WUFI® is shown exemplarily in Fig. 3 for a west facing wall with ETICS. During night there is no visible difference between measured and calculated surface temperatures. At noon the calculated course shows a delayed heating up. The reason for this behaviour is that WUFI® has to calculate the west radiation from direct and diffuse solar radiation. To do this, the diffuse radiation is assumed to be isotropic. In

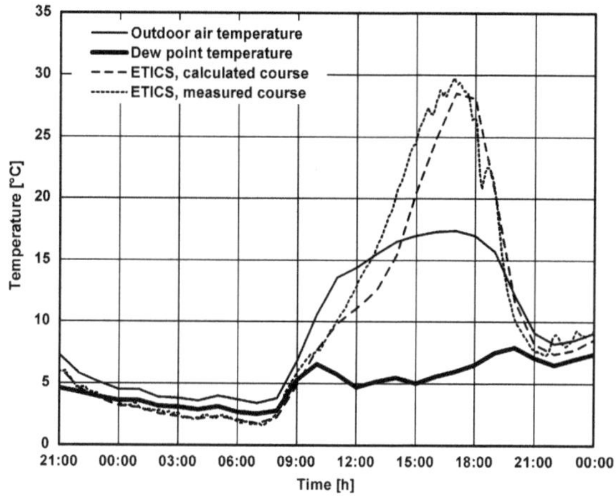

Fig. 3 Comparison of measured and calculated courses of surface temperature for a west facing construction with ETICS for a clear night and the following day in September. The courses of outside air temperature and dew point temperature are also shown

reality there is an increase of the diffuse radiation towards the sun. Around noon the calculated west radiation is lower than actual measured values and therefore the calculated surface temperatures are lower compared to the measured values. Surface temperatures near (or below) the dew point temperature can only be observed at night time or in the early morning hours. Consequently WUFI® can calculate the surface temperatures for different variants to determine the periods of condensation and the degree of cooling below dew point temperature.

4.2 Influence of Construction Type and Orientation

Spring and especially autumn are the most critical times of year for microbial growth, since in winter it is mostly too cold and in summer mostly too hot and dry. Figure 4 shows the summarised hours below dew point in autumn for both constructions and two orientations. The difference between the construction with ETICS and the monolithic one is obvious. The monolithic wall gets more condensation when oriented eastwards compared to the wall with ETICS. In this case the stored solar energy of the morning sun is partially released until dawn, while on the western side the energy can be stored during night time. With ETICS this effect is not dominant because of its low thermal capacity. In total the amount of condensation is obviously higher with ETICS compared to the monolithic wall of aerated concrete. This corresponds with observations concerning biological growth on facades in practice. Therefore, further investigation concentrate on ETICS.

4.3 Computations of Prevention Strategies

There are principally two different physical approaches to reduce condensation on the external surfaces of ETICS. By enhancing the thermal storage capacity close to

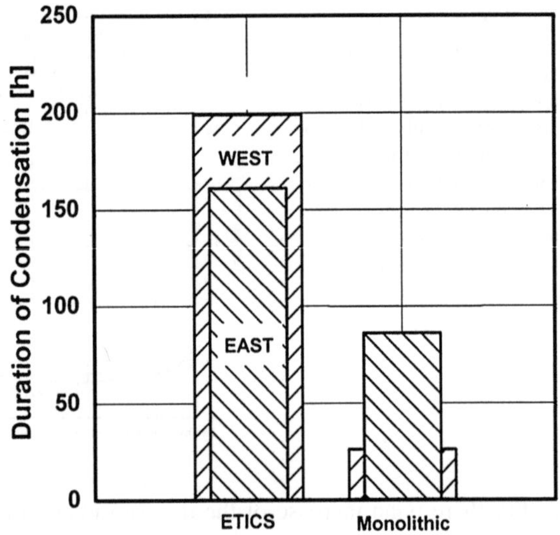

Fig. 4 Comparison of the summarised hours below dew point in autumn for a construction with ETICS (Polystyrene with $\rho = 30$ kg/m^3) and a monolithic one made of aerated concrete ($\rho = 600$ kg/m^3) for two orientations "east" and "west"

the exterior surface, it may be possible to store enough heat from the daily solar warming of the building component to avoid the lowering of temperature below the dew point temperature. In addition, the daily warming of the façade can be enhanced by applying paint with a higher short wave absorption factor. The decrease of temperature during night can be reduced by application of paint with a reduced long-wave emissivity (Low-E paints). A possible criterion to assess the results and the risk of growth is the duration of condensation and the intensity of the dew point temperature undercut. An other approach to reduce water on the external surfaces of ETICS is the improvement of the hygric material properties of the rendering, which could lead to a lower amount of moisture on the surface.

4.3.1 Influence of the Thickness of the Plaster

A possibility to reduce the time of wetness on the surface is to use a thicker layer of plaster. Normally a system with 3 mm plaster is applied. The first comparison to the standard construction is with a test façade with a thick plaster (10 mm). For façades with ETICS the thickness of the outer plaster layer should have an influence on the course of surface temperature because of the correspondent heat capacity. In Fig. 5 the courses of the surface temperatures during a night in September are shown.

The temperature of the façade with a thick plaster decreases below the dew point temperature approximately 1h later than the temperature of the standard construction with 3 mm plaster. But in the morning when the dew point temperature and the surface temperatures rise the temperature of the façade with thick plaster needs more time to gets above the dew point temperature. The effect of a thicker plaster is not much more than a time shift of circa 1h in comparison to the standard construction.

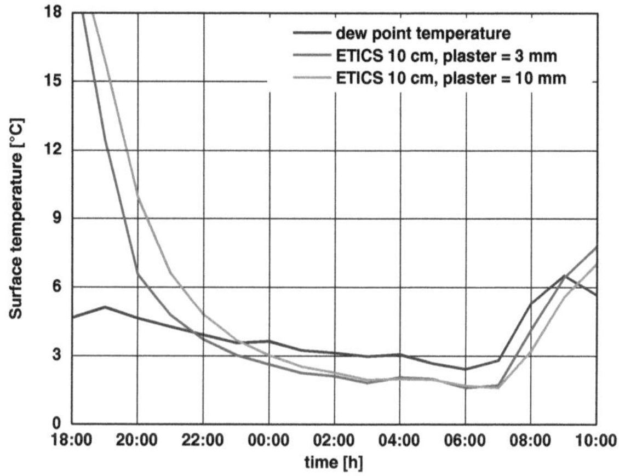

Fig. 5 Courses of the surface temperature of two western oriented façades with 10 cm ETICS and different thicknesses of plaster (3 mm and 10 mm) during a night in September

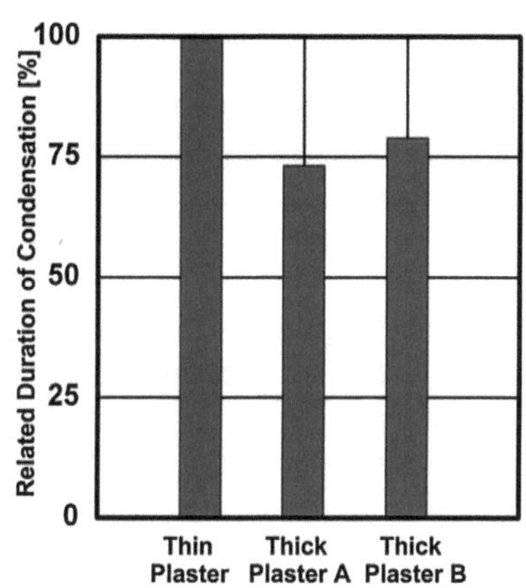

Fig. 6 Duration of condensation for two different thick plasters with a maximum thickness of 20 mm and a high density in comparison of a normal thin plaster

Figure 6 shows the duration of condensation for two different thick plasters with a maximum thickness of 20 mm and a high density in comparison of a normal thin plaster with a thickness of about 4 mm. This gives an imagination of the upper limits of thick plaster for prevention of night time condensation on the surface.

4.3.2 Influence of Irradiative Properties of the Surface

To analyze the influence of irradiative properties of the surface calculations have been conducted with a dark colour of the rendering instead of a white one (radiation coefficient of 0.6 instead of 0.4) and with a lowered infrared emissivity (long wave radiation coefficient of 0.6 instead of 0.9). As expected the surface temperature of the dark coloured plaster is reaching the maximum temperature during the day as shown in Fig. 7. A new promising way to decrease the condensation on

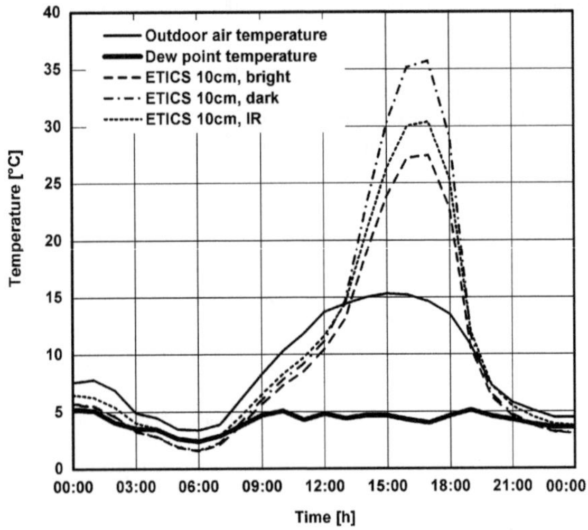

Fig. 7 Courses of surface temperatures for west facing walls with ETICS and different surface properties. Data for a sunny summer day (13th September). The curves for outdoor air temperature and dew point temperature are shown additionally

ETICS is the application of Low-E paintings. The positive effect of such paint has been investigated by hygrothermal calculations, too. The version with a bright plaster but low IR-emissivity is getting warmer than the standard case. This means that even during the day each surface has a non neglectable energy loss due to long wave emission.

During nighttimes the thermal irradiation causes a cooling down of the wall (Fig. 8). The surface temperature of the dark coloured facade, which shows the highest temperatures during the day, is cooling down below dew point temperature for the same duration as the white one. This shows that the thermal mass of the rendering on top of the insulation is too low to store enough solar energy for the prevention of condensation. The surface temperature of the ETICS with low IR-emissivity remains above dew point temperature instead.

Figure 9 shows the accumulated duration of condensation for walls with ETICS in the main growth period autumn for different surface properties. It is evident that a dark colour only causes a slight improvement, which means a reduction of condensation of about 5 %. The advantage of a dark colour lies in the higher surface temperatures after sunrise, which enhances the drying of the rendering. The good performance for ETICS with low-E coating compared to conventional paints demonstrates that these can be a promising possibility to reduce the risk of algal growth on ETICS.

4.3.3 Using Phase Change Materials

Another possibility to prevent condensation on the surface of the façade is the use of PCMs. The idea is, that while the surface of the building is loosing energy due to long wave radiation the temperature of the PCM is staying constant as long as it is changing its phase. If this temperature lies above the dewpoint temperature of

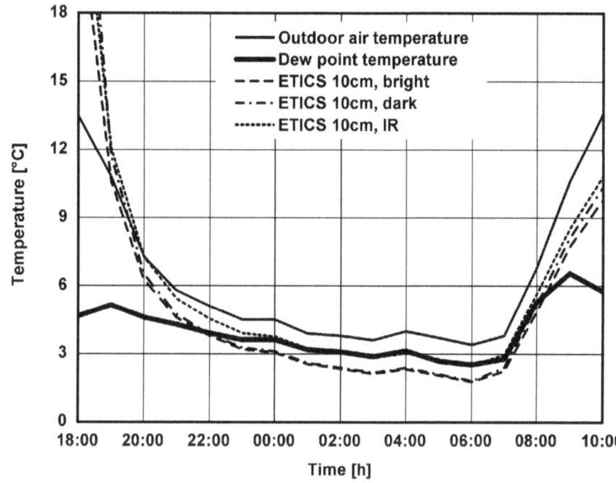

Fig. 8 Courses of surface temperatures for west facing walls with ETICS and paints with different spectral properties. Data for a clear night (13th, 14th September). The curves for outdoor air temperature and dew point temperature are shown additionally

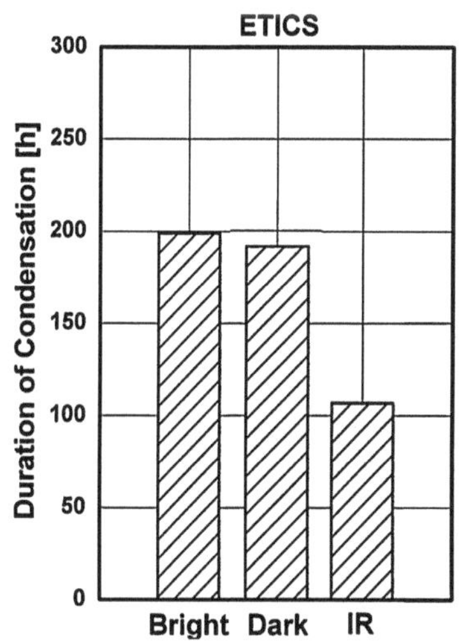

Fig. 9 Accumulated duration of condensation for walls with ETICS in the main growth period autumn for different surface properties

the ambient air no condensation will occur. For the computation of this effect a PCM-layer behind the plaster with a thickness of 5 mm is assumed. For the PCM paraffine may be used, which is available in a wide range of melting temperatures. Therefore, the melting temperature may be adopted to optimize the influence on the duration of condensation. Here two types of PCM were taken as a basis, differing in their physical properties (see Table 1).

Figure 10 shows the results for a variety of temperature ranges using the outer climate conditions of Holzkirchen, Germany. It is obvious that the choice of the phase change temperature has an important influence on the effectiveness of the PCM.

Figure 11 displays the course of the surface temperature for a typical night in the middle of September for a wall with a standard ETICS and a wall with PCM

Table 1 Material properties of the implemented PCM types

Material property	Unit	PCM type 1	PCM type 1
Bulk density	(kg/m^3)	900	1500
Thermal conductivity	(W/m K)	0.2	1.0
Heat capacity	(J/kg K)	2100	1500
Latent heat	(kJ/kg K)	100	110

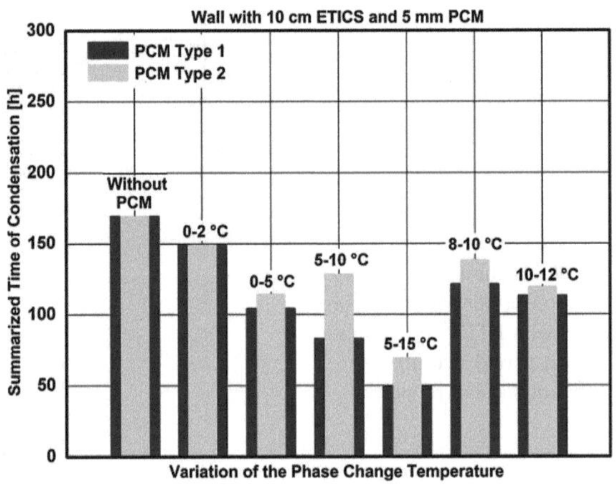

Fig. 10 Computed results for a variation of temperature ranges using the outer climate conditions of Holzkirchen

which is not optimized (phase change at about 15 °C) and an optimized one (phasen change at 5–12 °C). It is clearly visible that with an optimized PCM the wall has remarkably higher surface temperature during nocturnal hours than the conventional ETICS. However, in the morning hours a delay of the temperature rise can be observed. This is caused by the phase change from solid to liquid during this period. These results in condensation in the morning, but because of the rising outside temperature the condensate will dry out fastly. The course of the surface temperature of the wall with PCM, whose melting point isn't optimized shows nearly the same behaviour during night-time as the standard ETICS. Therefore, only small improvements concerning condensation will occur.

In Fig. 12 the surface temperatures below the dew point temperature of the ambient air are cumulated for the standard ETICS and the system with an optimized layer of PCM. With PCM the amount of condensation is more than a factor of two below that of common ETICS.

4.3.4 Combination of PCM and IR-Effect

It is reasonable to combine both measures, the use of PCM and the application of paint with reduced long wave emissivity. The impact of the PCM and the IR-effect

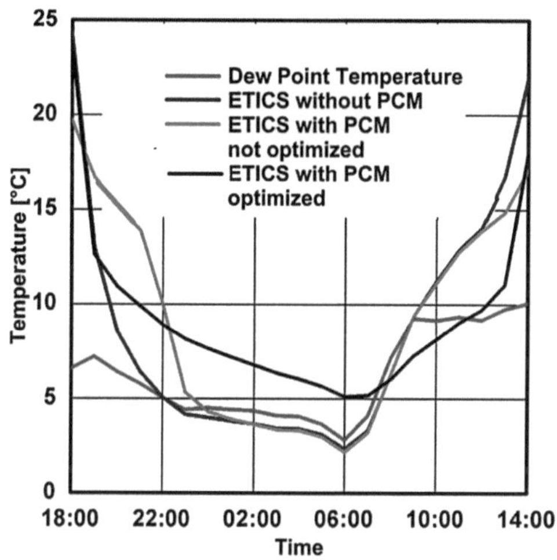

Fig. 11 Course of the surface temperature for a typical night in the middle of September for a wall with a Standard—ETICS and a wall with PCM which is not optimized (phase change about 15 °C and an optimized one (phasen change at 5–12 °C)

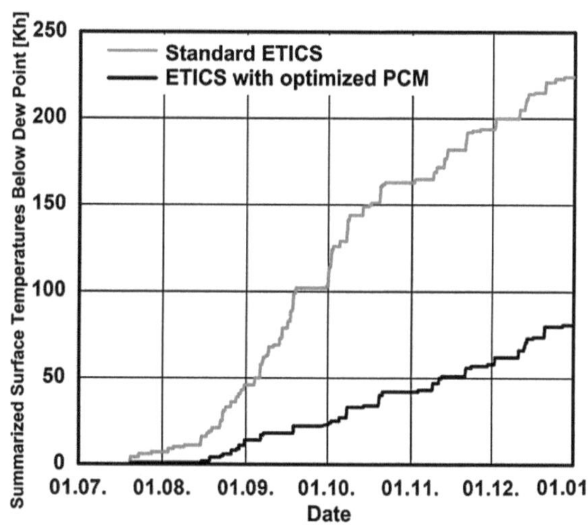

Fig. 12 Cumulated surface temperatures below the dew point temperature of the ambient air for the standard ETICS and the system with an optimized layer of PCM

of the materials is demonstrated in Fig. 13. In the afternoon the temperatures of the three façades decrease in a different way depending on the properties of the surface and the material. The standard construction (blue line) with a grey coating reaches the lowest surface temperature. The temperature is below the dew point temperature during the whole night from 5 P.M. to 5 A.M. The influence of the PCM (black line) is resulting in a higher temperature, also for the façade with PCM and IR-effect (green line). The temperature of the façade with PCM is below the dew point temperature only for short periods between 6–8 P.M. and 2.30–5 A.M. At that time the stored latent heat is released. In the morning the temperature of the standard construction increases faster because the façades with PCM need more energy to warm up. In contrast to this the façade with PCM and IR-painting shows no problems with condensation almost during the whole night.

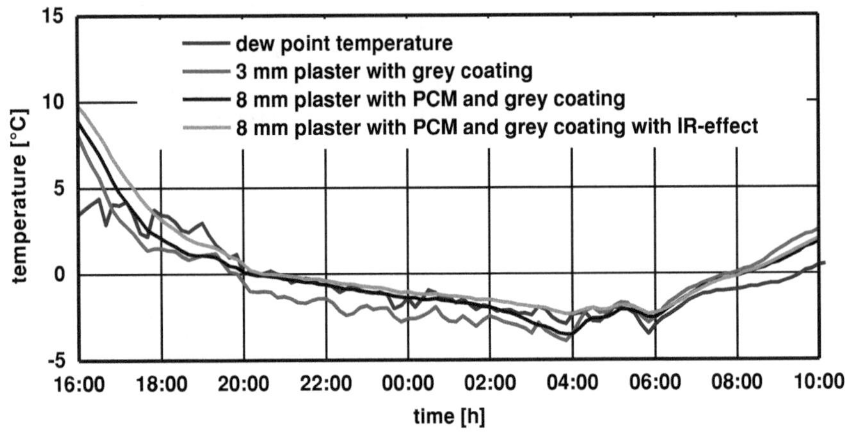

Fig. 13 Courses of the surface temperature of three northern oriented façades with 10 cm ETICS and different grey coatings with and without IR-effect and with PCM in a night in November

4.3.5 Comparison of Measures to Reduce the Amount of Condensation

Figure 14 shows the expected effects of the different measures. It can be clearly seen that the application of darker façade paint does not yield any noteworthy result with regard to avoiding condensation. Nevertheless, a tinted paint can be advantageous, since it conceals slight growth and entails higher temperatures from time to time, which may be lethal to micro-organisms. A thicker external plaster with a high thermal mass can reduce dew point temperature undercut by a maximum of 20 %, IR paint by almost 30 %, and a latent thermal storage layer even by 70 %. The combination of both measures (PCM + IR) can further reduce the duration of dew point temperature undercut, as an extreme case. It must, however, be taken into consideration that in case of applying (PCMs) the desired effect will only occur, if the melting point of the PCM is adjusted to the climate boundary conditions, which is expressed by the additional term "optimized" in Fig. 14.

4.3.6 Improvement of the Hygric Material Properties

The most important influential criterion is the moisture content on the surface, caused by the climate and building physical parameters. Physical properties of the surface are important besides surface orientation, thermal insulation and the thermal storage capacity of building components. As it is obvious that already small differences in the mean or long-term level of humidity cause visible differences in growth, a control of the hygric surface parameters could lead to a profound reduction of the biological infestation. In this context, it remains to be clarified how far hydrophoby of the surface, providing lower water absorption of the external render, will have the commonly expected positive effect of a reduction

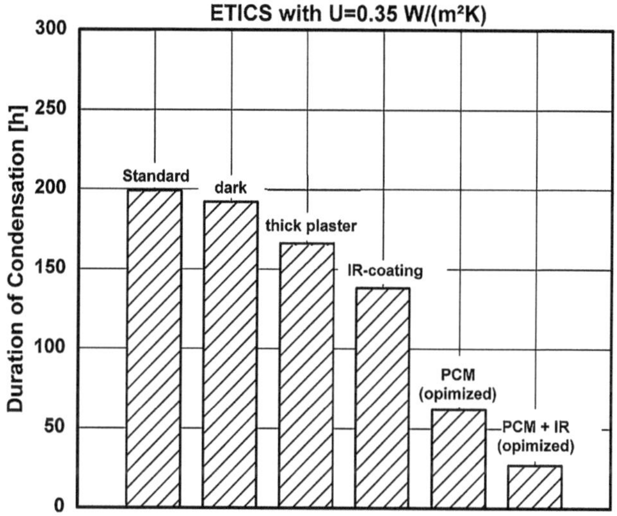

Fig. 14 Effects of various measures on the duration of the dew point temperature undercut in the period from September to October

of pollution and microbial growth. It is also possible that a certain degree of absorbency of the substrate will help to avoid microbial growth by deviating condensation water from the surface deeper into the building components so that it is no longer available for micro-organisms.

With a recently developed model for the hygrothermal calculation of the moisture on the surface (Krus and Rösler 2006) the influence of the hygric properties of the rendering has been investigated. Figure 15 shows the amount of water on the external façade for an ultra-hydrophobic paint and a dispersion silicate coating, accumulated during the main growth period of micro-organisms in autumn. It is apparent that the dispersion silicate coating provides only half of the amount of water for micro-organisms.

5 Field Tests

Field tests have been conducted on two test façades to verify the calculated results. The test façade with northern orientation is a standard ETICS with different top paints, while the facade with Western orientation has a monolithic construction. Beneath the outer surface of the walls the course of the surface temperature has been measured to indicate the duration and intensity of condensation by comparing this temperature with the measured outdoor dew point temperature. Figure 16 shows a photographic view of the western oriented (right) and the northern oriented (left) façades being tested.

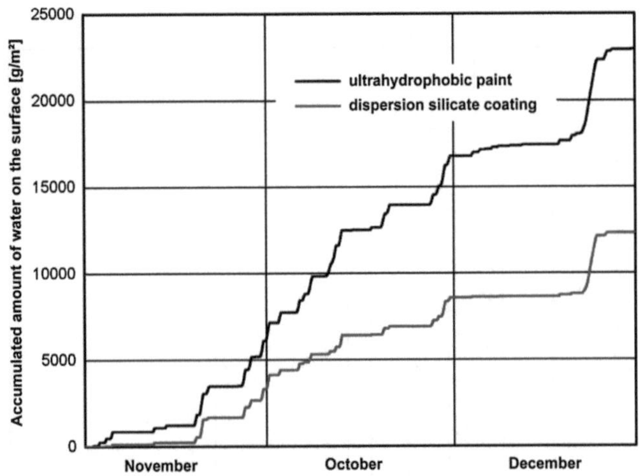

Fig. 15 Amount of water on the external façade for an ultra hydrophobic paint and a dispersion silicate coating, accumulated during the main growth period of micro-organisms in autumn

Fig. 16 Photos with the view of the western oriented (*right*) and the northern oriented (*left*) façade

5.1 Influence of Construction Type

The measured results of massive walls clearly show in comparison to an ETIC system, that the impact of condensation is lower on monolithic systems (see Fig. 17) as they yield to shorter condensation duration. As the calculations already showed, the grey colour only yielded a negligible improvement.

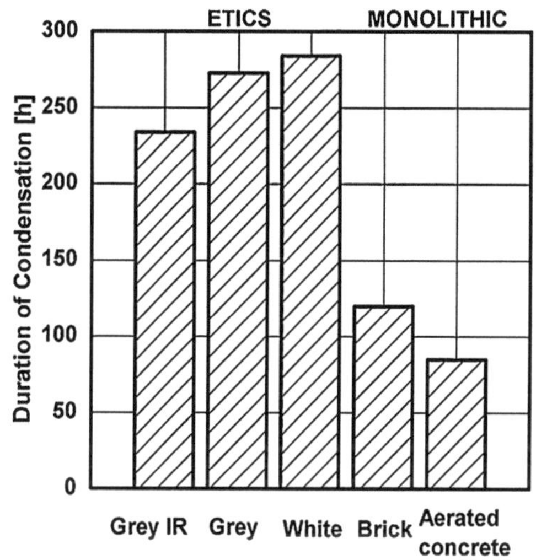

Fig. 17 Summarized condensation times for various wall constructions and paints for the most important period of growth in autumn

5.2 *Use of PCM and Low-E Paint*

The advantage of innovative materials, e.g. IR paints, PCM plasters and the combination of PCM and IR paint can be proven by measurement (see Fig. 18, compared with results from the northern façade shown in Fig. 16). Since temperatures are too low for microbial growth in winter, autumn (as general observations confirm) is the main period of microbial growth. With the measured condensation times observed for this period of time, the effectiveness of PCM is clearly visible in the first autumn, although measurements could be started not before October. In comparison to a white thin plaster duration of condensation was reduced by approximately 30 % for thick plaster with PCM additives. When comparing with the calculated results for PCM in Fig. 14 it must be taken into consideration that less than half the quantity of PCM was added so that the measured duration of condensation tends to show good compliance with the calculation results. The façade with IR paint, which only achieved an emissivity value of 0.74 instead of 0.6 in contrast to the calculations, shows a decrease of the condensation times by approx. 15 %. As expected, the combination of PCM plaster and IR paint achieves a reduction of the exterior condensation of almost 50 %, clearly a higher reduction than the results with PCM additive and IR paint alone.

Besides the duration and quantity of condensation the actual amount of surface water is decisive to assess the risk of microbial growth, since only the water on the surface of a façade is available for micro-organisms. Therefore, the surface humidity on the western façade without PCM plaster was determined by dabbing the water on top of the surface. Figure 19 shows the results of a morning after a clear night. Almost double the amount of water was measured on the surface of a standard ETICS with white paint than on the surface of a white thick plaster. It must, however, be taken into consideration that these results are dependent on the hygric material properties of the plaster and paint to a very large extent. More

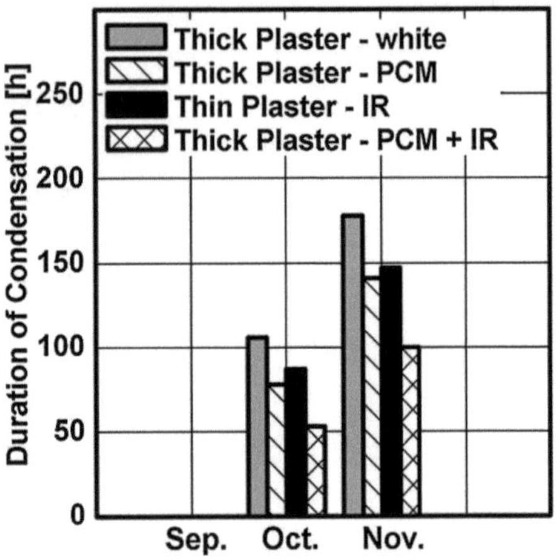

Fig. 18 Duration of condensation for the ETICS without and with PCM and IR paint for the first autumn of the measurement period

condensation is penetrating through the thick plaster beneath the surface. The best result is achieved by the surface with IR paint (variant with an emission coefficient of approximately 0.6). Almost no water could be measured on the surface this morning. Nevertheless, these IR paints have to be optimised for practical use, because up to now all tested available IR-paints are getting a metallic glance after weathering (see Fig. 20). This is due to the special binding agent, which has to be transparent for long wave radiation but isn't robust enough for the outdoor use. Here aluminium flakes, which were used to reduce long-wave emission, appeared on the surface. The measurements, however, show that it would be beneficial to further develop IR paints to adjust weathering resistance to present requirements.

The measurements on the northern façade of Fig. 16 were continued. Figure 21 displays the results for the successive two periods in autumn. Surprisingly, the façade with PCM additives shows no or only a slight reduction of the condensation times in contrast to the first period in autumn. The reason is not the stability but the melting temperature of the PCM used, which amounts to 6 °C. Whereas the mean ambient air temperature was 5 °C in the autumn of the first year of investigations, it rose to approx. 10 °C in the following years. Therefore, no latent heat could be released. The façade with IR paint, however, shows a continuous and constant lowering of condensation times. An interesting aspect is the fact that the combination of PCM plaster and IR paint shows a clearly higher reduction of the condensation production during the total period of measurements than the IR paint alone, although the PCM additive alone did not yield any improvement from time to time.

5.3 *Influence of Hygric Material Properties*

To determine the temporal progression of surface moisture under natural boundary conditions, the method of dabbing with a subsequent weighing of the fibrous web was used at the western-oriented external façade with different thermal insulation

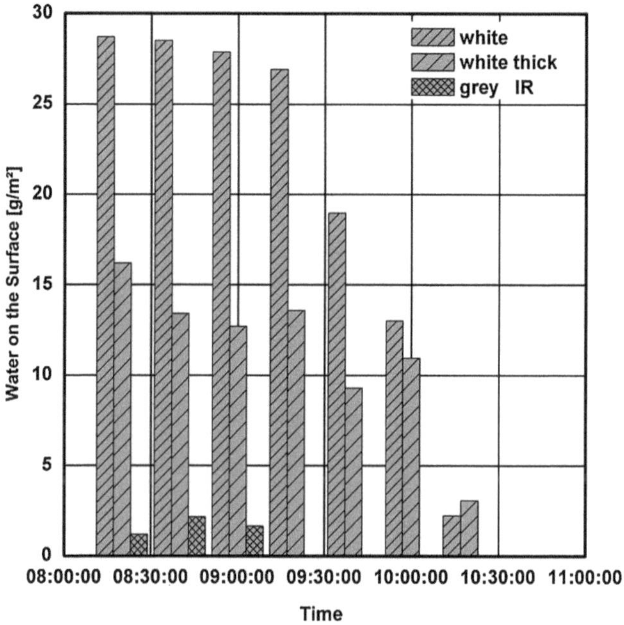

Fig. 19 Amount of surface humidity of western oriented walls with various ETICS at different times in the morning

Fig. 20 Photographic view of the surface with an IR-paint before (*left side*) and after a weathering period of 12 months (*right side*)

composite systems and coatings. The results are presented in Fig. 22 for a 2h interval in September. At the first measuring time at 8:00 A.M., the amount of condensation water on the thick and thin plaster system with mineral coating was remarkably lower than on the thin plaster system with silicone resin paint and ultra-hydrophobic coating. The amount of condensation water is still growing in

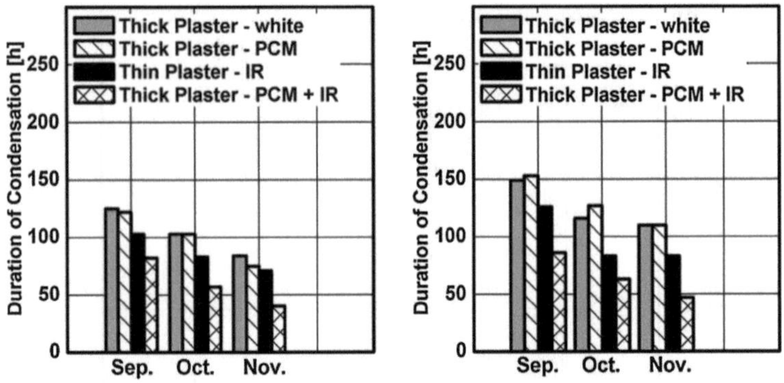

Fig. 21 Duration of condensation on the surface of the ETICS without and with PCM and IR paint of the second autumn (*left*) and the third autumn (*right*) of the period of measurements

Fig. 22 Gravimetric determination of the surface moisture on a test wall with different coatings

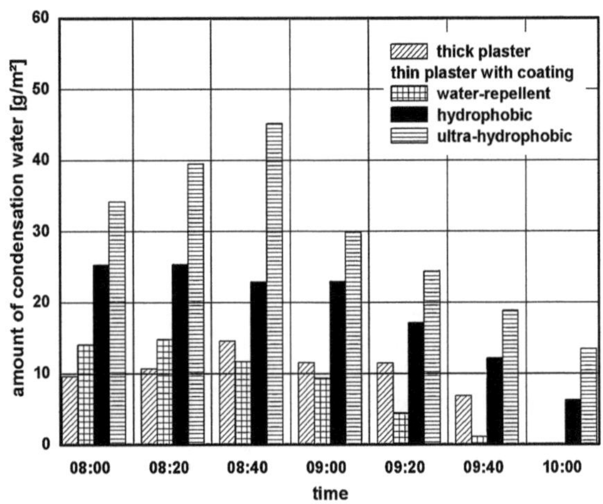

the beginning concerning the wall sections with the ultra-hydrophobic coating, whereas it remains almost constant with other surfaces. The wall sections with thick and thin plasters with water-repellent coating dry in the temporal progression, while the wall with the thick plaster dries out more slowly than the wall with the thin plaster. At 10:00 A.M., both walls are already dry at the surface, whereas on the silicone resin paints and especially on the ultra-hydrophobic coating some condensation water is still left.

6 Summary

In the last decades, the level of thermal insulation was considerably increased especially for new buildings. The improvement of the standard of thermal insulation resulted in a higher risk of microbial growth on external façades by mould and algae. The essential criterion for the risk of microbial growth on façades is the availability of sufficient humidity. Condensation at night caused by nocturnal long-wave emission is of special importance, since it is the only explanation for the observed increase in microbial growth on the northern façade with only little driving rain. To assess the risk of microbial growth condensation on the surface is thus a good criterion. The direct comparison of monolithic walls, e.g. made of aerated concrete or bricks, with walls with ETICS shows that the latter are more at risk. But to improve the thermal insulation of already existing buildings the application of ETICS is the most practicable way in most cases. Therefore, possible solutions against microbial growth are investigated for these systems. Calculations as well as field test measurements were conducted for this purpose. If we consider the problem of undesired microbial growth on façades, we must notice that the majority of ETICS is without any damage. Only a minor percentage shows microbial growth. For this reason the authors suggest that already a reduction of condensation duration by 25 % could avoid most problems. However, there are always special circumstances, for example a shaded wall in the vicinity of a forest or water, where the application of chemical agents is inevitable.

Principally there are three different physical approaches to reduce condensation on the external surfaces of ETICS. By enhancing the thermal storage capacity close to the exterior surface, it may be possible to store enough heat from the daily solar warming of the building component to avoid the lowering of temperature below the dew point temperature. In addition, the daily warming of the façade can be enhanced by applying paint with a higher short wave absorption factor. The drop in temperature at night can be reduced by means of applying paint with a reduced long-wave emissivity (IR paints). A possible criterion to assess the results and the risk of growth is the duration of condensation and the intensity of the dew point temperature undercut.

Another way to reduce nocturnal condensation is to use latent heat effects by adding so-called PCMs in external façades. Paraffin for example, which is available with various melting point ranges, can serve as PCM. The optimization of the temperature for the phase change is essential to reduce the condensation by applying PCM. The problem, however, is its application. The optimization of the phase change point is calculated by applying test reference years. Yet, the differences of the climate conditions in successive years are so immense that not even this kind of optimized PCM can be continuously effective. This example, however, also shows that calculations alone do not reflect practical reality despite correct calculation results, since parameters are neglected from time to time, which can only be revealed by outdoor testing, in this case over a period of several years. Future investigations will show whether long-term calculations over a period of at least ten measured years will enable a better optimization of the PCM. Another way may be the use of a PCM-material with a wide range of melting temperature, which means a mixture of different paraffin's. Since the amount of PCM which may be added to the plaster is too small a supplementary layer of PCM behind the rendering should be taken into account.

In case of IR paints the emission coefficient for long-wave radiation can be reduced from more than 90 % to approx. 60 %. Higher maximum temperatures during the day as well as a reduction of condensation are achieved by reduced thermal emission. This is the method, which is the simplest approach for already existing buildings. Unfortunately, IR paints, which are sufficiently resistant to weathering, are not yet available.

The most essential criterion for the risk of microbial growth on façades is the availability of a sufficient amount of water. According to the present state of knowledge, it is obvious that not the amount of condensation water is decisive for the risk of microbial growth but the amount of water on the external surface, which is available for micro-organisms in the initial phase of growth. Absorbent substrates could offer remarkable advantages in this respect. It has to be mentioned that with the hygric optimisation it is not a matter of reaching systems with high water absorption coefficients. The nocturnal condensation is a very slow process with a low amount of moisture. This means that it is sufficient to raise the water absorption coefficient slightly from the extreme low A value of common hydrophobic systems with 0.0005 kg/m^2 s to values around or something above 0.002 kg/m^2 s. Therefore it will be still a water-repellent coating system. These positive results led to the fact, that in the meantime numerous rendering systems with higher absorptivity have been developed for the European market.

References

Blaich, J.: Außenwände mit Wärmedämm-Verbundsystem, Algen- und Pilzbewuchs (Facades with ETICS, algae and mould growth). Deutsches Architektenblatt **31**(H.10), S.1393–1394 (1999)

Denffer von, D., Ziegler, H., Ehrendorfer, F., Bresinsky, A.: Lehrbuch der Bo¬tanik für Hochschulen (Textbook of the botany for universities). In: Begründet von, E., Noll, F.S., Schenk, H., Schimper, A.F.W. (eds.) 32. Auflage/neubearbeitet von Dietrich von Denffer, Hubert Ziegler, Friedrich Ehrendor¬fer, Andreas Bresinsky. Gustav Fischer, New York (1983)

Hofbauer, W., Breuer, K., Sedlbauer, K.: Was wächst auf unse¬ren Fassaden? (What is growing on top of our facades?) Teil I: Algen, Flechten, Moose, Farne. Bauphysik. **25**, H. 6 (2003)

Hofbauer, W., Fitz, C., Krus, M., Sedlbauer, K., Breuer, K.: Prognoseverfahren zum biologischen Befall durch Algen, Pilze und Flechten an Bauteiloberflächen auf Basis bauphysikalischer und mikrobieller Untersuchungen (Assesment procedure for microbial growth of algae and mould on the basis of physical and biological investigations). Hrsg.: Fraunhofer-Institut für Bauphysik–IBP-, Holzkirchen; Bauforschung für die Praxis Band 77 (2006). IRB-Verlag. ISBN 978-3-8167-7102-9

Krus, M., Sedlbauer, K.: Instationärer Feuchtegehalt an Außen¬oberflächen und seine Auswirkungen auf Mikroorga¬nismen (Transient moisture on outside surfaces and ist effect on micro organisms). Tagungsbeitrag zur IBK-Bau-Fachtagung 288 Bau¬schäden durch Schimmelpilze und Algen, 27 Feb., S 5/1–5/15 Berlin (2003)

Krus, M., Rösler, D.: New model for the hygrothermal calculation of condensate on the external building surface. Third International Building Physic Conference, Montreal (2006)

Krus, M.; Fitz, C.: Vergleich unterschiedlicher Messverfahren zur Erfassung der Oberflächenfeuchte von Putzfassaden. (Comparison of different measurement techniques for the determination of surface moisture on renderings). Berichtsband zum Workshop 2007 Innovative Feuchtemessung in Forschung und Praxis, S. 119–126. Aedificatio-Verlag. (2007) ISBN 978-3-931681-91-3

Künzel, H.M.: Simultaneous heat and moisture transport in building components. One- and two-dimensional calculation using simple parameters. PhD Thesis, University of Stuttgart (1994)
Künzel, H.M., Schmidt, Th., Holm, A.: Exterior surface temperature of different wall constructions—comparison of numerical simulation and experiment. 11. Bauphysika¬lisches Symposium, Dresden (2002)
Scherer, S.: Anpassungen von Cyanobakterien in Wüsten (Adaptations of cyano bakteria in deserts). In: Hausmann, K., Kremer, B.P. (eds.) Extremophile: Mikroorganismen in ausgefallenen Lebensräumen, S. 179–193. VCH, Weinheim, Basel, New York, Cambridge, Tokyo (1993)
Sedlbauer, K.: Vorhersage von Schimmelpilzbildung auf und in Bauteilen (Prediction of Mould Growth on Top of and Inside Building Parts). Thesis, University of Stuttgart (2001)
Venzmer, H.: Grüne Fassaden nach der Instandsetzung durch WDVS? Nicht bestellt und dennoch frei Haus (Green Facades after renovation with ETICS? Not ordered but delivered freely). 3. Dahlberg-Kolloquium (2001)

Biological Defacement of External Thermal Insulation Composite Systems

Eva Barreira, Vasco Peixoto de Freitas and João M. P. Q. Delgado

Abstract External Thermal Insulation Composite Systems (ETICS) are nowadays often used in Europe. Despite its thermal advantages, low cost and ease of application, this system has serious problems of microbiological growth causing the cladding defacement. This chapter presents the results of a detailed experimental study that was carried out in order to assess the hygrothermal behaviour of façades covered with ETICS, namely to evaluate the influence of orientation on surface humidification, by external condensation and by wind-driven rain. A building located in the University of Porto campus, with its four façades facing the cardinal directions, was monitored during one year. The exterior surface parameters under study were temperature and wind-driven rain. The building interior conditions and the exterior climate were also measured. The tests results were compared with the results of the numerical simulation performed using a commercial hygrothermal model. The conclusions about the results of the experimental and numerical study, their comparison and all the difficulties found to achieve similar results are pointed. This chapter also presents a methodology to assess the risk of biological growth, based on the definition of indices, which combine the effect of surface condensation, wind-driven rain and drying process, three of the most prevalent parameters influencing the surface moisture content. The proposed indices were calculated using data collected during the "in situ" test campaign, which provided information about the exterior climate conditions, the

E. Barreira (✉) · V. P. de Freitas · J. M. P. Q. Delgado
LFC—Building Physics Laboratory, Civil Engineering Department, Faculty of Engineering, University of Porto, 4200-465, Porto, Portugal
e-mail: barreira@fe.up.pt

V. P. de Freitas
e-mail: vpfreita@fe.up.pt

J. M. P. Q. Delgado
e-mail: jdelgado@fe.up.pt

V. P. de Freitas and J. M. P. Q. Delgado (eds.), *Hygrothermal Behavior, Building Pathology and Durability*, Building Pathology and Rehabilitation 1,
DOI: 10.1007/978-3-642-31158-1_2,

Fig. 1 ETICS defacement due to biological growth (bulking located in porto - Portugal)

surface temperature and wind-driven rain on four façades covered with ETICS facing the cardinal directions. The indices were compared with the results of the surface relative humidity measured simultaneously that allowed the validation of the methodology. An example of the practical use of this methodology is also presented in this chapter, with the definition of hazard classes of biological defacement of façades covered with ETICS located in the Portuguese territory.

Keywords ETICS · Biological defacement · Hygrothermal behaviour · "In situ" measurements · Numerical simulation

1 Introduction

External Thermal Insulation Composite Systems (ETICS) are often used in Europe since de 1970s, both in new buildings and in retrofitting. Despite its thermal advantages, low cost and ease of application, this system has serious problems of microbiological growth, causing the cladding defacement. Although no changes occur in the thermal and mechanical performance of the system, biological defacement has an enormous aesthetic impact, which gathers the building's dwellers disapproval and restricts the full implementation of this technology (see Fig. 1). Recent studies allowed to understand the physical phenomenon causing the ETICS defacement and made it possible to develop mathematical models simulating its performance. However, no simple process has yet been developed to predict the risk of ETICS defacement, which may be used by designers and by the building industry.

Studies carried out during the last decade (Blaich 1999; Kunzel and Sedlbauer 2001; Becker 2003; Venzmer et al. 2008) pointed that microbiological growth is due to high values of surface moisture content, which results from the combined effect of four parameters: surface condensation, wind-driven rain (WDR), drying process and properties of the exterior layer.

Exterior surface condensation occurs mostly during the night, when the exterior surface temperature is lower than the dew point temperature, as a result of long wave radiation exchange between the surface and the atmosphere. During clear nights, the atmosphere's emitted radiation decreases considerably and the radiation emitted by the surface is greater than the one that reaches the surface, causing a loss of radiation towards the sky. This negative balance on the surface is maintained until heat transport by convection and by conduction compensate for the loss by radiation (Holm et al. 2004).

WDR results from the combination of rain and wind. Its intensity depends on the building geometry and environment topology, position on the building façade, intensity of rainfall through a horizontal plane, wind speed and wind direction. The most used semi-empirical model says that WDR intensity is proportional to the product of the wind speed normal to the façade and the intensity of rainfall through a horizontal plane (Hoppestad 1955; Nore et al. 2007).

The drying process allows the evaporation of the liquid water accumulated on the surface due to the surface condensation and WDR. Evaporation from the wet surface occurs whenever the saturation pressure at the surface is greater than the vapour pressure of the ambient air (Hagentoft 2001). If the drying process is not sufficiently fast, the surface moisture content remains high for long periods and increases the risk of microbiological growth (Krus et al. 2006).

Several authors refer to the importance of the properties of the exterior layer in the biological growth. Concerning the thermal properties, emissivity is referred as the most relevant as it rules the emission of radiation by the surface. For lower values of emissivity the heat loss during the night reduces and the surface temperature increases. The risk of surface condensation is therefore lower (Krus et al. 2006). Also some hygric properties are referred as having capital influence on the surface moisture content, namely water absorption of the external coating. Lower water absorption allows longer availability of liquid water on the surface as less water is absorbed by the plaster system following WDR and surface condensation (Blaich 1999; Becker 2003; Venzmer et al. 2008). Barberousse et al. (2006) assessed the influence of the coating roughness in the biological growth and stated that it could promote or restrict the adhesion process of the spores and the liquid water retention on the surface.

Although no statistical studies are available, façades covered with ETICS facing North and West are more prone to biological growth than the ones facing East and South, which rarely have any problem. This great influence of orientation in the façades defacement, although known by the researchers and by the producers/appliers of the system, has never been deeply studied. Its real impact in the surface humidification by condensation and WDR is therefore unknown.

Fig. 2 Building under study—located in University of Porto campus

2 Experimental Study

2.1 Setting Up the Test

To assess the influence of orientation in the surface humidification an "in situ" test campaign was carried out during one year (March 2009–February 2010). Instruments were set up on the façades covered with ETICS of a building located in UP campus (see Figs. 2 and 3), whose walls face the cardinal directions. T-type thermocouples were used to assess surface temperature and WDR gauges measured the rain reaching each façade. Surface relative humidity was measured using humidity and temperature probes. They were coupled to a data acquisition system, collecting data every 10 min. Information regarding the accuracy and calibration of the devices is given by Barreira (2010). Black dots on Figure 3 mark the position of the devices on each façade.

Climatic parameters were collected every 10 min by the weather station located near the building under study (see Fig. 3) and the hygrothermal conditions inside the building were also measured. The annual averages of the outdoor and indoor climate are presented in Table 1. Detailed information about the weather station (instruments used and its accuracy) is given by Barreira (2010).

2.2 Results

2.2.1 Surface Temperature

Figure 4 shows the surface temperature of the façades under study and the dew point temperature, during two days of May 2009. During the night, surface temperatures were quite constant and very similar. Even so, the South façade had the

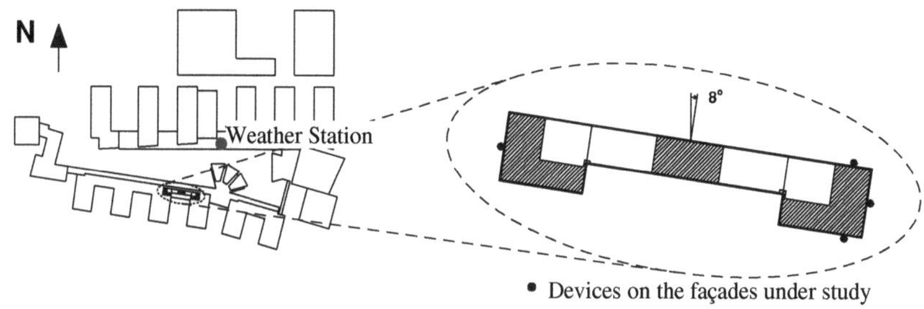

Fig. 3 Layout with the devices and weather station locations

Table 1 Indoor and outdoor climate during the test campaign (annual average)

Climatic parameter	Used devices	Outdoor	Indoor
Temperature	Pt100 sensor	15.4 °C	19.9 °C
Relative humidity	Hygrometer sensor	72 %	67.5 %
Global radiation emitted by the sun	Pyranometer	254 W/m^2	–
Radiation emitted by the sky	Pyrgeometer	335 W/m^2	–
Wind velocity/direction	Anemometer/vane	1.4 m/s/170°	–
Rain (accumulated)	Rain gauge	874 mm	–

higher temperature and the West façade the lower one. The temperature on the North façade was not the lowest as it was expected.

During daylight, when the sky is clear, the differences between the surface temperatures are more obvious. The East and South façades achieved higher values during the morning and the West façade at the end of the afternoon. The temperature of the North façade was the lowest. The peak at daybreak is related with direct solar radiation incident on the façade. When the sky is cloudy, the differences during daylight are smaller and almost independent from orientation (see 17th May). Figure 5 displays the average, maximum and minimum surface temperatures measured for the period under study.

2.2.2 Exterior Surface Condensation

When surface temperature drops below dew point temperature (T_{dp}), calculated as a function of temperature and relative humidity of the ambient air, condensation will occur. The difference between T_{dp} and surface temperature may be called Condensation Potential in degrees (CPd), which implies condensation for positive values. The accumulated value of positive CPd during a period of time allows estimating the amount of water vapour available to condensate and points to the risk of condensation.

Figure 6 shows that hourly CPd was always lower than 2.5 °C and, on average, was around 0.4 °C. Although the major differences had occurred during the cold seasons, on average, there was no significant variation of CPd along the year. The

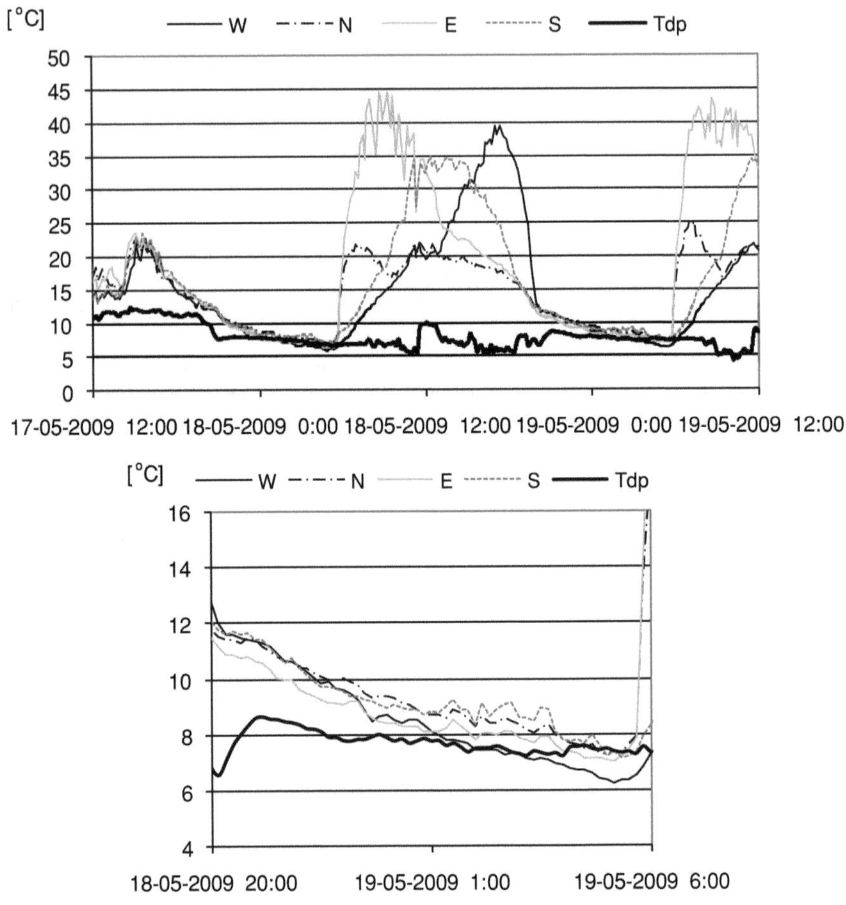

Fig. 4 Surface temperatures and dew point temperature during May 2009

accumulated values of CPd (only the positive temperature differences were considered, the negative values were taken as equal to zero) point to higher risk of condensation during November, December and April. During July and August the risk of condensation was also significant (Fig. 7), which is related with Porto climatic conditions. During summer nights high relative humidity and mild temperature lead to high T_{dp}, worsen the risk of condensation. Considering the annual accumulated CPd, the West façade presented higher risk of condensation, followed by the East, North and South façades.

2.2.3 Wind-Driven Rain

WDR measurements show that, in generally, the South façade was more exposed to rain, followed by the West, East and North façades. The amount of WDR was

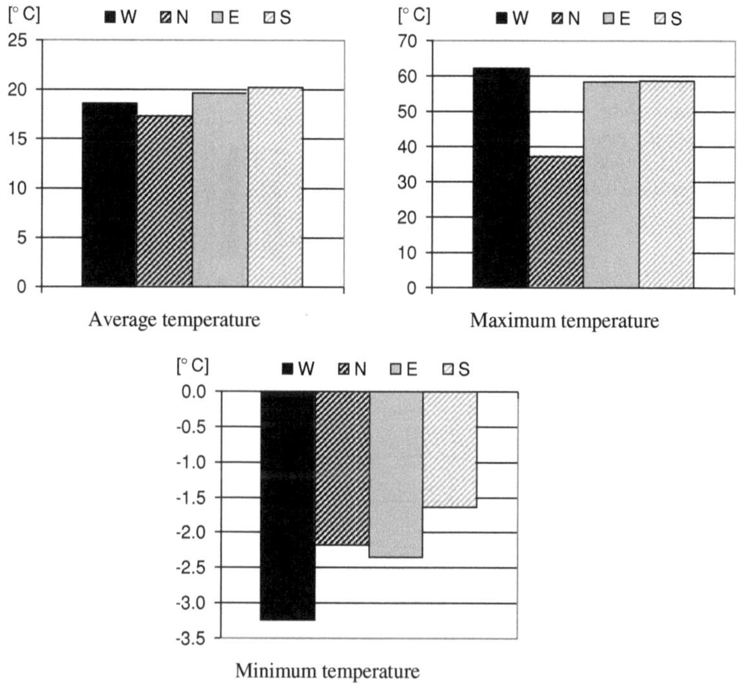

Fig. 5 Average, maximum and minimum temperatures measured during the test campaign

higher during autumn and winter. It must be said that the amount of rain reaching the façade may correspond to maximum values as the WDR gauges were placed near the top and vertical edges of the façades. The values of WDR for the West façade during January and February 2010 are not available due to problems in the gauge (see Fig. 8).

2.3 Discussion of the Results

During the night, surface temperature does not differ much with orientation (see Fig. 4). The small differences are due to direct solar radiation incident on the façades and on its surrounding during the day, which vary with orientation. The North façade did not had the lowest surface temperature during the night (see Figs. 4 and 5).That may be related with the rotation of the façades, about 8 °Clockwise from North (see Fig. 3), and with the exposure to the wind of the North façade, which is aligned with the main façade of the building unlike the other façades (see Fig. 2).

The comparison between the risk of surface condensation (see Fig. 7) and the defacement of the façades under study (North and West façades present more colonization by microorganism than the East and South façades that do not have

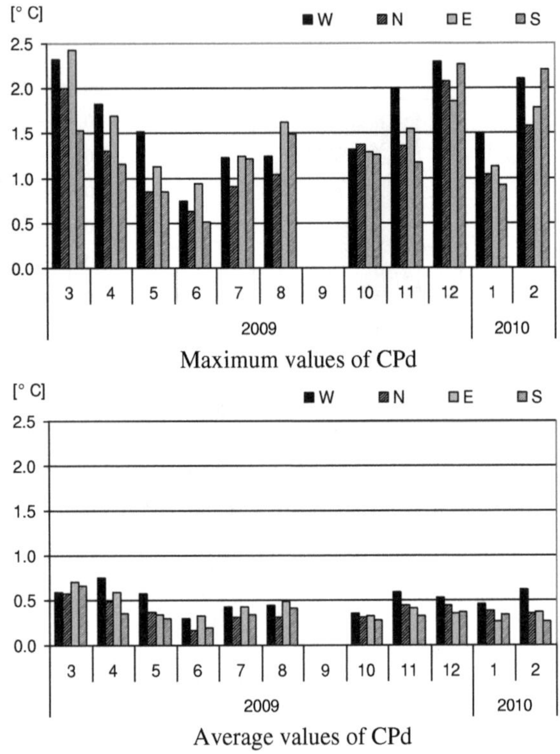

Fig. 6 Maximum and average values of CPd obtained during the test campaign. (In September 2009 there were some problems with the data acquisition system and the measurements were lost)

any biofilm on its surface) points to the influence of the drying process in the surface water content and biological growth. Lower temperatures on the North façade during daylight (see Fig. 5) may restrict the evaporation of the condensated water and increase the risk of biological colonization. The West façade had the highest amount of condensation and the second highest amount of WDR. The surface temperature reached during day may not be enough to decrease the water content on the surface and restrict biological growth. The South façade, although having less risk of condensation, had the highest amount of WDR, more than the double measured in the West façade. As it is not defaced, the drying process may have a key role in biological growth on ETICS (Barreira and Freitas 2011a).

In Porto, external surface condensation may be more preponderant to biological growth than WDR, as it occurred during all the year at equivalent intensity (see Fig. 7). During spring and summer, the rain did not reach the façades for longer periods (see Fig. 8) pointing to a smaller influence in the surface water content. Of course, the water run-off along the surface and, specially, its accumulation may increase the influence of WDR in the risk of microorganisms' colonization (Barreira and Freitas 2011b).

In Sects. 4 and 5 is presented a methodology to assess the risk of biological growth on ETICS. It points to the major influence of the drying process based on the drying potential for each orientation. It also indicates that surface condensation

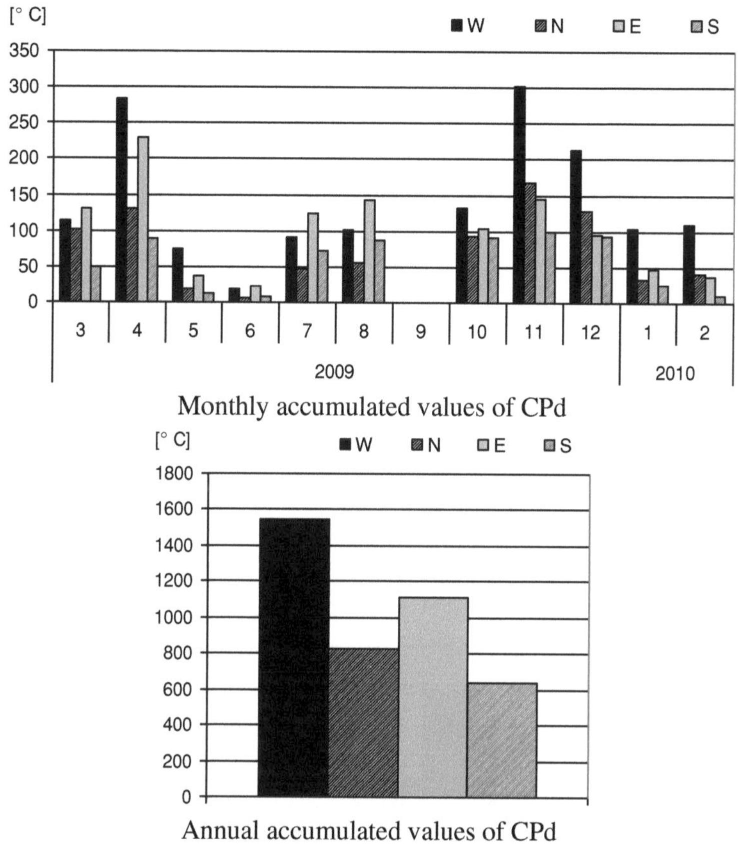

Fig. 7 Monthly and annual accumulated values of CPd obtained during the test campaign. (In September 2009 there were some problems with the data acquisition system and the measurements were lost)

has higher impact on biological growth than WDR, as long as water run-off is not taken into account.

3 Numerical Simulation

3.1 Parameters of the Numerical Simulation

In building physics hygrothermal models are widely used to simulate the coupled transport of heat and moisture. However, only a few allow simulating accurately the exterior surface temperature during the night on a façade covered with ETICS.

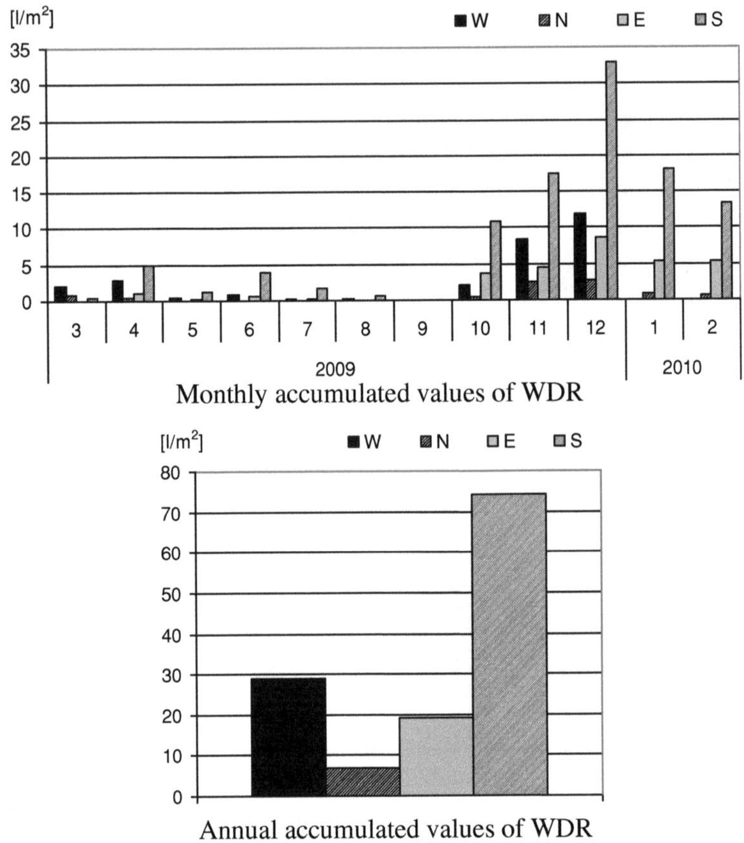

Fig. 8 Monthly and annual accumulated values of WDR obtained during the test campaign. (In September 2009 there were some problems with the data acquisition system and the measurements were lost)

The low thermal capacity of the external rendering and its thermal decoupling emphasises the influence of radiation, which implies including a routine to calculate explicitly the radiative balance on the surface. This radiant balance is affected by the building's radiation, the sky's radiation, terrestrial surface's radiation and solar radiation (Delgado et al. 2010).

The need for heat, air and moisture (HAM) models to better understand the phenomena behind external surface condensation demands for their wide validation. For that reason, from the existing hygrothermal models, it was selected a commercial one (Künzel 1995, WUFI 2008, Kehrer and Schmidt 2008) and the results obtained by numerical simulation were compared with the ones measured during the experimental campaign. The simulations were run using the wall from Fig. 9. The thermal and hygric properties of the materials used in each layer were selected from the software database, as an approach to the existing wall. The

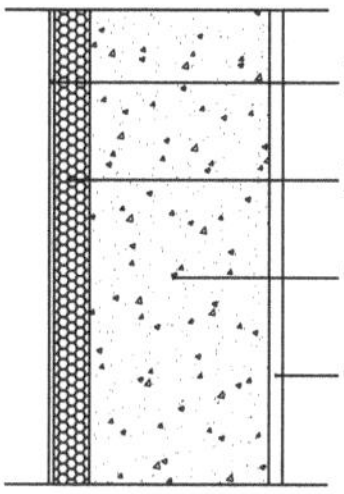

Fig. 9 Wall used for the numerical simulation

hourly climatic data used was measured by the weather station (see Fig. 3) and indoor temperature and relative humidity were measured using probes. Table 1 shows the annual averages of the climatic parameters measured, although for the numerical simulation hourly values were used.

3.2 Discussion of the Results

3.2.1 Surface Temperature

Figure 10 shows the hourly average of surface temperature on the West and South façades, measured and calculated, during one day of November 2009. During the night there is a good agreement between the measured and the calculated values for the West façade. For the South façade, a better agreement was achieved after increasing 4 % the sky's radiation measured (*Sim_a*). This adjustment may be related with the effect of the ground in the façade surface temperature during the night. The ground, with high thermal capacity, absorbs solar radiation during the day that is released during the night as long wave radiation. The amount of radiation absorbed by the ground is not the same for each cardinal direction due to shading caused by the building itself.

During daylight, the differences between the measured and the calculated values are strongly marked. That may be related with the rotation of the façades (see Fig. 3) that could not be considered in the simulation due to limitations of the model. The solar radiation values measured by the weather station were lower than the ones presented in the literature, which may have affected the simulation results, reducing the calculated surface temperature during daylight. The calculation of the sun position in the horizon, performed by the software to calculate the direct solar radiation on the façade, may also have introduced some errors. The differences obtained during a sunny daylight are smaller than the ones obtained with cloudy sky, as the temperatures measured and calculated are much closer (see Fig. 11).

3.2.2 Exterior Surface Condensation

The good agreement obtained for the surface temperature during the night induced similar measured and calculated values for the PCd (see Fig. 12), during the colder months. During the warmer ones the influence of the sun was higher, which implied more divergent results (see Fig. 13).

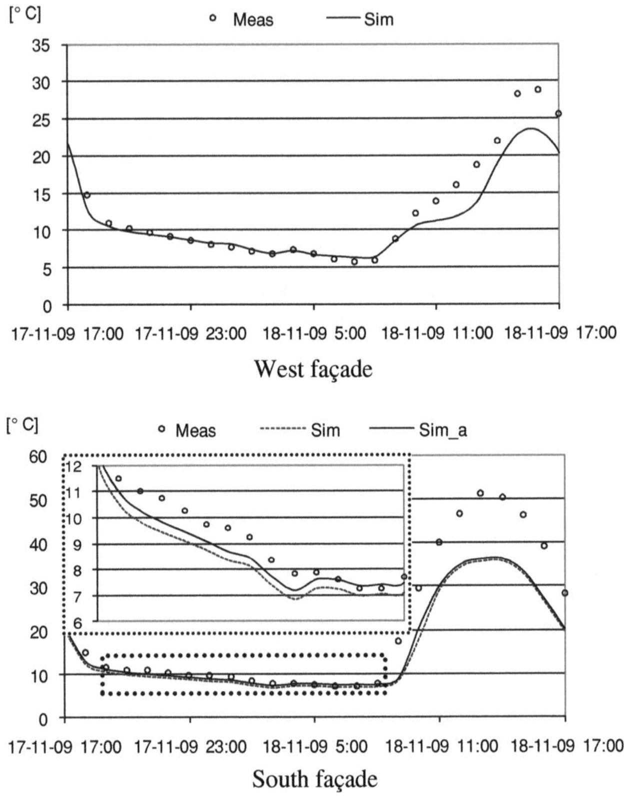

Fig. 10 Surface temperature, measured and calculated, during a clear day in November 2009

3.2.3 Wind-Driven Rain

The agreement between measured and simulated values of WDR (see Fig. 14) was only achieved after adjusting the driving rain coefficients used by the software to estimate the rain load on the building component (see Table 2). This adjustment was probably necessary due to the rotation of the façades and the different wind conditions near the building and near the weather station, as the weather station is located between two buildings and the building under study is exposed to open wind (see Fig. 3).

4 Hygrothermal Model to Assess the Risk of Biological Growth

The knowledge gathered during the last decade made it possible to develop mathematical models simulating ETICS performance, considering different conditions of use (Kunzel et al. 2002). However, no simple process has yet been

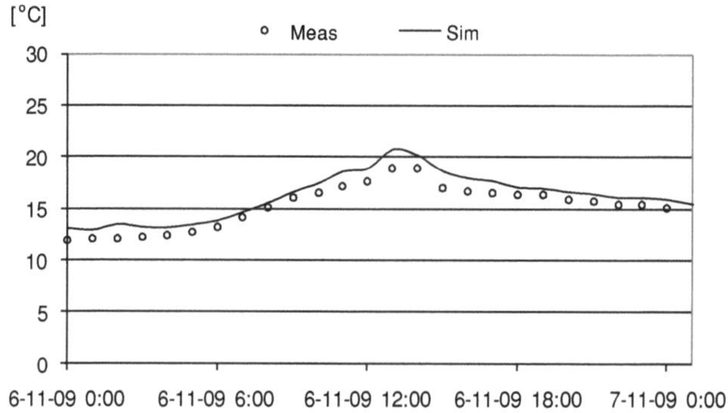

West façade

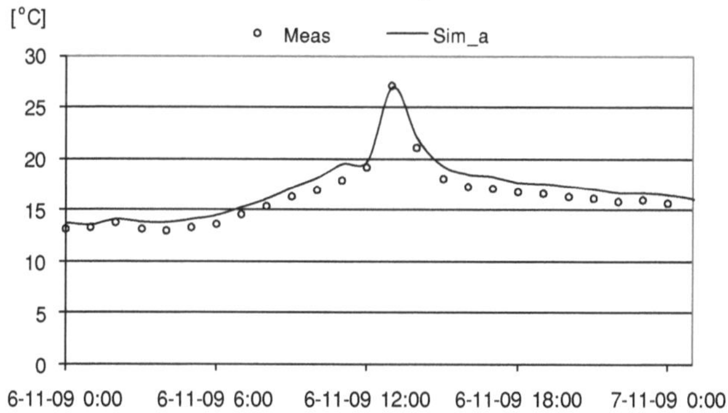

South façade

Fig. 11 Surface temperature, measured and calculated, during a cloudy day in November 2009

developed to predict the risk of ETICS defacement. Considering the work already done for mold growing on interior finishes (Adan 1994 Viitanen 1996, Sedlbauer 2001), the relative humidity of the surface or the time of wetness obtained from the surface humidity, might be the better criterion for assessing the risk of biological growth. Nevertheless, the comparison of simulated values with the results of "in situ" tests performed on a façade covered with ETICS showed that there is no go agreement between the simulated and the measured values of the relative humidity, especially when WDR is taken into account (Barreira 2010).

The use of surface relative humidity as a criterion for the risk assessment is, therefore, restricted. The next sections present an alternative methodology to assess the risk of biological growth, based on the definition of indices, which combine the effect of surface condensation, wind-driven rain and drying process, three of the most prevalent parameters influencing the surface moisture content.

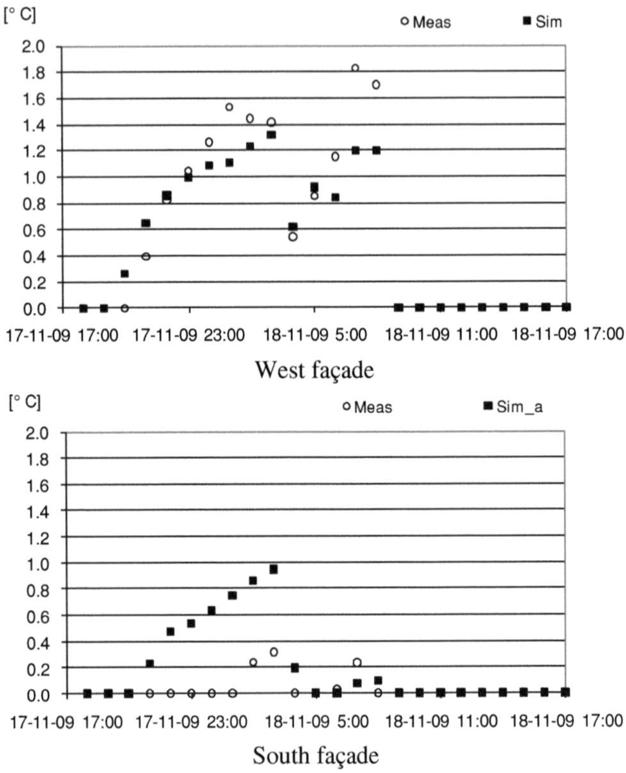

Fig. 12 PCd, measured and calculated, during a clear day in November 2009

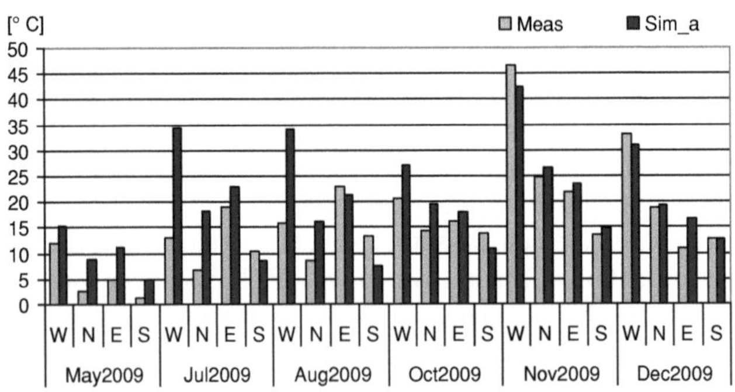

Fig. 13 Monthly accumulated PCd, measured and calculated

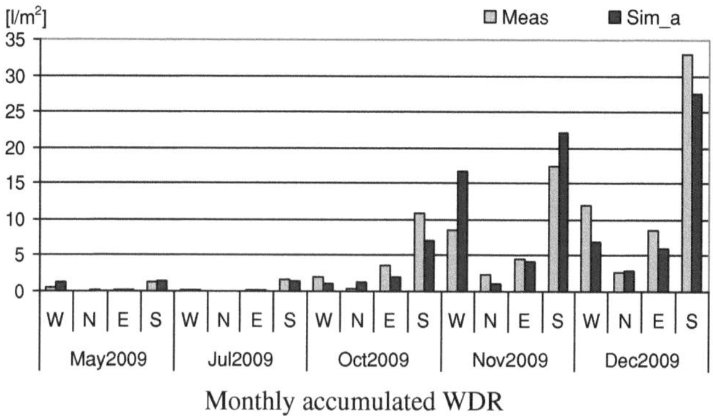

Fig. 14 Monthly accumulated WDR, measured and calculated

Table 2 Driving rain coefficients used for the simulation

Orientation	West	North	East	South
Driving rain coefficients (before adjustment)	0.2	0.2	0.2	0.2
Driving rain coefficients (after adjustment)	0.3	0.2	0.02	0.07

4.1 Assessing Exterior Surface Condensation

Exterior surface condensation can be analyzed using psychrometry principles. When water vapour partial pressure of the air is greater than the water vapour saturation pressure at the surface, condensation will occur (Hagentoft 2001). According to Zheng et al. (2004) the difference between the water vapour partial pressure in the air ($P_v(air)$, in Pa) and the water vapour saturation pressure on the surface ($P_{sat}(surface)$, in Pa) may be called Condensation Potential (*CP*, in Pa), which implies condensation for positive values. *CP* can be understood as the amount of water vapour that is available to condensate.

$$CP = P_v(air) - P_{sat}(surface) \quad (1)$$

The same author refers that to evaluate the amount of condensation, both positive *CP* and its lasted time should be considered. The product of positive *CP* ($CP_{(>0)}$, in Pa) by its lasted time ($\Delta t_{CP(>0)}$, in h) may be called Condensation Potential Equivalent (*CPE*, in Pa.h) and allows estimating the amount of condensated water. To estimate the risk of condensation for a certain period of time *CPE* must be accumulated in time (CPE_a).

$$CPE = CP_{(>0)} \cdot \Delta t_{CP_{(>0)}} \quad (2)$$

4.2 Assessing WDR

The humidification of a façade due to *WDR* may be assessed, for a certain period of time, through the *WDR* Potential Equivalent ($WDRPE_a$, in kg/m^2), which is obtained by integrating in time the intensity of *WDR* (in kg/(m^2 s). $WDRPE_a$ shall be multiplied by 100 in order to reach values that are comparable with CPE_a values.

$$WDRPE_a = 100 \cdot \int_0^t WDR\,dt \tag{3}$$

4.3 Assessing the Drying Process

As condensation also the drying capacity of a wet surface can be analyzed using psychrometry principles (Hagentoft 2001). By analogy, it is possible to establish the concept of Drying Potential (*DP*, in Pa), as being the difference between the water vapour saturation pressure on the surface ($P_{sat}(surface)$, in Pa) and the water vapour partial pressure in the air ($P_v(air)$, in Pa), which implies evaporation for positive values. *DP* can be understood as the amount of water vapour transferred to the air, considering that the surface remains permanently wet.

$$DP = P_{sat}(surface) - P_v(air) \tag{4}$$

To evaluate the maximum ability to dry out, the product of positive *DP* ($DP_{(>0)}$, in Pa) by its lasted time ($\Delta t_{DP(>0)}$, in h) shall be considered and may be called Drying Potential Equivalent (*DPE*, in Pa.h). To estimate this ability for a certain period of time, *DPE* must be accumulated in time (DPE_a).

$$DPE = DP_{(>0)} \cdot \Delta t_{DP_{(>0)}} \tag{5}$$

It must be stated that DPE_a is not useful as a parameter for modeling the real drying capacity of a wet surface, as it is not permanently saturated. After some time, the liquid water evaporates and the vapour pressure at the surface depends not only on the surface temperature, but also on its relative humidity. However, to avoid the use of relative humidity and to simplify the parameters used in the drying process assessment, DPE_a can be employed as an overvalued drying capacity.

4.4 The Model BIO.MOD

The model BIO.MOD defines three indices: BIO.MOD1 that relates surface humidification by condensation (CPE_a) with the maximum drying capacity (DPE_a); BIO.MOD2 that relates surface humidification due to WDR ($WDRPE_a$)

with the maximum drying capacity (DPE_a); and BIO.MOD3 that relates surface humidification by condensation (CPE_a) and due to WDR ($WDRPE_a$) with the maximum drying capacity (DPE_a) and is the one to be considered when analyzing the risk of ETICS defacement due to high amount of surface moisture content (Barreira and Freitas, 2011b).

$$BIO.MOD1 = \frac{CPE_a}{DPE_a} \cdot 10^3 \tag{6}$$

$$BIO.MOD2 = \frac{WDRPE_a}{DPE_a} \cdot 10^3 \tag{7}$$

$$BIO.MOD3 = \frac{CPE_a + WDRPE_a}{DPE_a} \cdot 10^3 \tag{8}$$

5 Validation of the Model BIO.MOD

Using the data collected during the "in situ" campaign (see Sect. 2.1), CPE_a, $WDRPE_a$, DPE_a and the three indices of BIO.MOD were calculated in annual bases, as sketched in Figs. 15 and 16.

Figure 17 shows that index BIO.MOD3, which combines surface condensation with the effect of WDR and assesses the risk of defacement due to biological growth, follows the same trend as the accumulated hours of the surface saturation (relative humidity equal to 100%), measured simultaneously, and allowed the validation of the model. On the other hand, there is also a good relation between the façades defacement per orientation and the annual values of *BIO.MOD3*, as the West and North façades present more colonization by microorganism than the East and South façades that do not have any biofilm on its surface.

The risk of surface humidification due to condensation (*BIO.MOD1*) is greater for the West façade, followed by the North, East and South façades. It depends mostly on the drying process, which is quite clear for the North façade that did not had high values for surface condensation (CPE_a) but had the lowest drying capacity (DPE_a) and, consequently, the highest risk of humidification by condensation (*BIO.MOD1*). The East façade, with one of the highest amount of condensation (CPE_a), did not present very high risk of humidification by condensation (*BIO.MOD1*) due to the drying capacity (DPE_a).

The risk of surface humidification due to WDR (*BIO.MOD2*) is greater for the South façade, followed by the West, East and North façades. The risk is related with the intensity of WDR ($WDRPE_a$) on each façade and there is no clear influence of the drying capacity (DPE_a), although it had slightly reduced the risk in the South façade.

The risk of defacement due to biological growth (*BIO.MOD3*) is strongly dependent on surface condensations (*BIO.MOD1*), as both indices follow the same trend. The influence of the WDR is not very relevant although it slightly increases

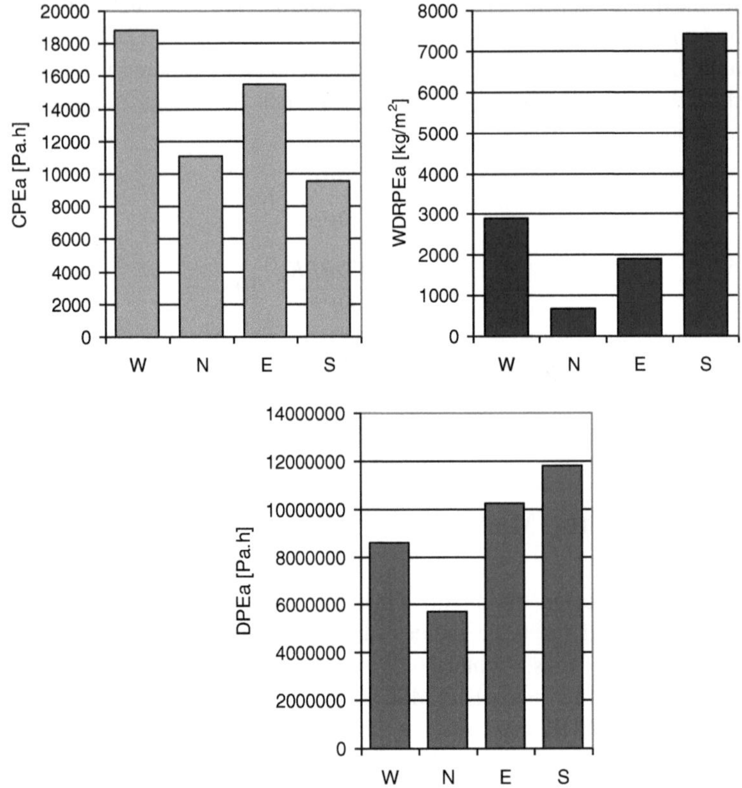

Fig. 15 CPE_a, $WDRPE_a$ and DPE_a per orientation (annual values)

the risk of biological growth. However, it must be stated that this model does not takes into account the water run-off along the surface neither its accumulation, which may increase the influence of WDR in the risk of microorganisms colonization.

6 Risk Map of Biological Defacement in Portugal

As an example of the practical use of the model BIO.MOD, it was defined hazard classes of biological defacement of façades covered with ETICS located in the Portuguese territory. A risk map was created using the index *BIO.MOD3*, on annual bases, calculated with the results of numerical simulation for the same wall, facing North and South, and located in main towns of Portugal. The simulation was performed considering generated climatic data for each town and the interior conditions were assumed to be identical.

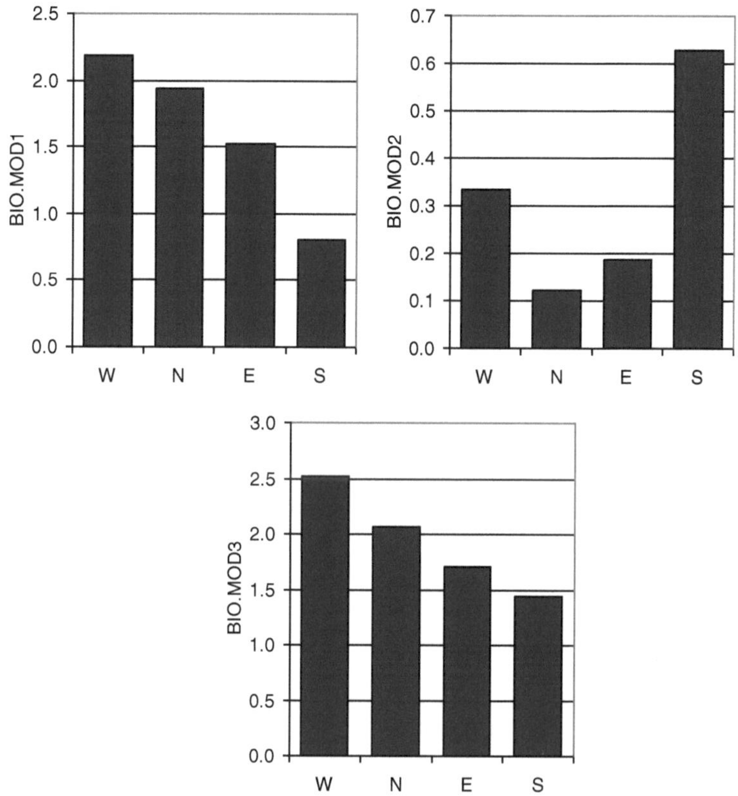

Fig. 16 Indices of BIO.MOD per orientation (annual values)

Fig. 17 Index *BIO.MOD3* and days with RH = 100 % per orientation (November 2009)

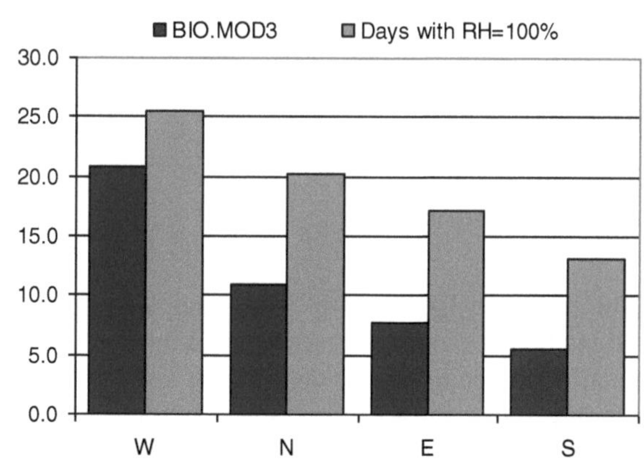

The hazard classes were established considering three classes of risk, having as top limit the highest value of *BIO.MOD3* obtained: Low Risk, considering values lower than 1.5, Medium Risk, for values between 1.5 and 3.0, and High Risk for

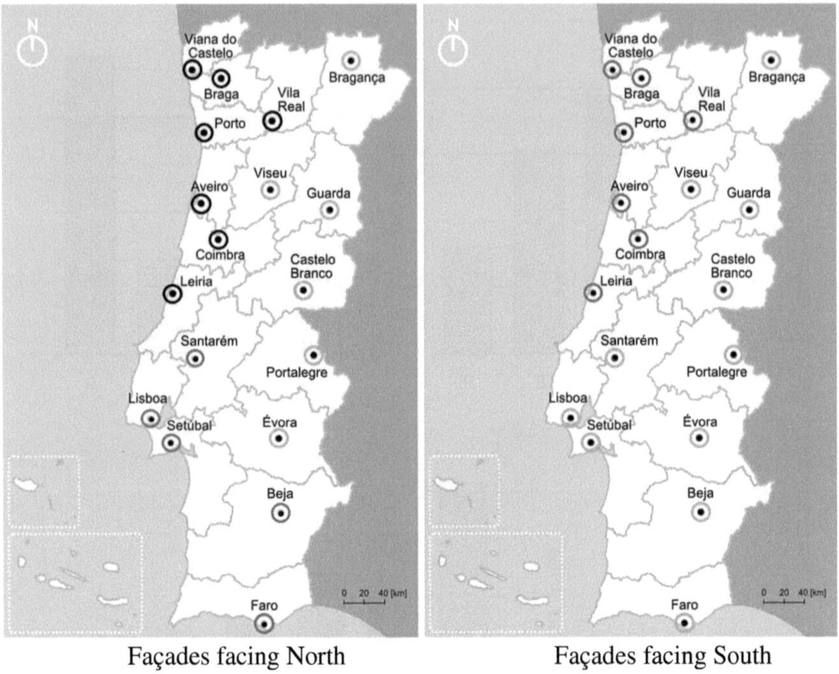

Fig. 18 Risk of biological defacement on façades facing North and South (*light grey* Low Risk,*grey* Medium Risk and *black* High Risk)

values higher than 3.0. Figure 18 displays the risk of biological defacement on façades facing North and South.

The results show that façades facing North have higher risk of defacement and the towns located in the North West of Portugal present more risk than the ones located in the North country side. In the South of Portugal, the risk is intermediate.

7 Conclusions

The results of the "in situ" tests campaign allowed understanding the effect of orientation in external surface temperature and WDR. During night, orientation had small influence in surface temperature and, consequently, in surface condensation. During the day, surface temperature varies considerably with orientation. Also WDR depends on the façade orientation. It was possible to point that biological growth depends not only on condensation and WDR, but also on the drying process.

The tests results allowed evaluating the performance of a commercial HAM model. Similar values of measured and calculated surface temperature were achieved for the night periods. During the day, due to the sun influence, the differences between the two were greater. The agreement for WDR was difficult to achieve. That may be related with the climatic data used as an input of the software.

It was developed and validated a methodology to assess the risk of biological defacement. This model is based on the definition of indices that relate humidification, by surface condensation and due to WDR, with the drying capacity. The main advantage of this methodology is related with the simplicity of the parameters used and with the ability of knowing the real load of surface condensation, WDR and drying capacity in the façade surface moisture content.

Using the model BIO.MOD it was possible to say that the drying process is the most relevant parameter and that surface condensation has more impact than WDR. It was also possible to establish a risk map for walls covered with ETICS located in Portuguese territory.

Acknowledgments The authors would like to thank the financial support of FCT—Fundação para a Ciência e Tecnologia that allowed the necessary conditions to carry out this study.

References

Adan, O.: On the fungal defacement of interior finishes. Ph.D. thesis, Eindhoven University of Technology, The Netherlands (1994)

Barberousse, H., Lombardo, R., Tell, G., Couté, A.: Factors involved in the colonisation of building façades by algae and cyanobacteria in France. Biofouling **22**(2), 69–77 (2006)

Barreira, E.: Biological defacement of façades covered with external thermal insulation systems due to hygrothermal behaviour. Ph.D. thesis, FEUP, Portugal (2010)

Barreira, E., Freitas, V.P.: Hygrothermal behaviour of ETICS—numerical and experimental study, NSB 2011. In: 9th Nordic Symposium on Building Physics, Tampere, Finland (2011a)

Barreira, E., Freitas, V.P.: Biological defacement of ETICS—a risk assessment methodology. In: 12DBMC International Conference on Durability of Building Materials and Components, Porto, Portugal (2011b)

Becker, R.: Patterned staining of rendered facades: hygro-thermal analysis as a means for diagnosis. J. Therm. Envelop. Build. Sci. **26**(4), 321–341 (2003)

Blaich, J.: La détérioration des bâtiments—analyse et prévention. EMPA, Suisse (1999)

Delgado, J., Freitas, V.P., Ramos, N., Barreira, E.: Numerical simulation of exterior condensations on façades: the undercooling phenomena. In : Proceedings of Thermal Performance of Exterior Envelopes of Whole Buildings XI, ASHRAE, Clearwater Beach, Florida, December 2010

Hagentoft, C.-E.: Introduction to Building Physics. Studentlitteratur, Sweden (2001)

Holm, A., Zillig, W., Kunzel, H.: Exterior surface temperature and humidity of walls—Comparison of experiment and numerical simulation. In: Proceedings of Performance of Exterior Envelopes of Whole Buildings IX, ASHRAE, Clearwater Beach, Florida, December 2004

Hoppestad, S.: Slagregn i Norge (Driving rain in Norway), Report No. 13, Norwegian Building Research Institute, Oslo, Norway 1955

Kehrer, M., Schmidt, T.: Radiation effects on exterior surfaces. In: Proceedings of 8th Symposium on Building Physics in the Nordic Countries, 2008, DTU, Copenhagen, Denmark, Vol. 1, pp. 207-212

Künzel, H.: Simultaneous heat and moisture transport in building components - One and two-dimensional calculation using simple parameters, IRB Verlag, Stuttgart, Germany, 1995

Kunzel, H., Sedlbauer, K.: Biological growth on stucco. In: Proceedings of Performance of Exterior Envelopes of Whole Buildings VIII: Integration of Building Envelopes, ASHRAE, Clearwater Beach, Florida, December 2001

Kunzel, H., Schmidt, Th., Holm, A.: Exterior surface temperature of different wall constructions—comparison of numerical simulation and experiment. In: Proceedings of 11th Symposium of Building Physics, Technische Universitat Dresden, Dresden, Germany 26–30 September 2002, vol. 1, pp. 441–449

Krus, M., Rosler, D. Sedlbauer, K.: New model for the hygrothermal calculation of condensate on the external building surface. In: Proceedings of Third International Building Physics Conference—Research in Building Physics and Building Engineering, Concordia University, Montreal, Canada 27–31 August 2006, pp. 329–333

Nore, K., Blocken, B., Jelle, B., Thue, J., Carmeliet, J.: A dataset of wind-driven rain measurements on a low-rise test building in Norway. Build. Environ. **42**, 2150–2165 (2007)

Venzmer, H., von Werder, J., Lesnych, N., Koss, L.: Algal defacement of façade materials—results of long term natural weathering tests obtained by new diagnostic tools. In: Proceedings of 8th Symposium on Building Physics in the Nordic Countries, DTU, Copenhagen, Denmark, 16–18 June 2008, vol. 1, pp. 277–284

Sedlbauer, K.: Prediction of mould manifestation on and in building parts, Thesis, University of Stuttgart, Germany, 2001

Viitanen, H.: Factors affecting the development of mould and brown rot decay in wooden material and wooden structures. Effect of humidity, temperature and exposure time, Dissertation, The Swedish University of Agricultural Sciences, Sweden, 1996

WUFI: WUFI Pro 4.2. Fraunhofer - IBP, Holzhirchen, Germany, 2008

Zheng, R., Janssens, A., Carmeliet, J., Bogaerts, W., Hens, H.: An evaluation of highly insulated cold zinc roofs in a moderate humid region—part I: hygrothermal performance. Constr. Build. Mater. **18**(1), 49–59 (2004)

Effectiveness and Durability of Biocides in Building Coatings: Biological Aspects

Nicole Krueger, Wolfgang Hofbauer, Martin Krus, Cornelia Fitz, Florian Mayer, Arnulf Melzer and Klaus Breuer

Abstract In an interdisciplinary project at the Fraunhofer-Institute for Building Physics in Holzkirchen, Germany about 350 biocide-equipped probes together with control specimens without biocide were exposed to real climatic conditions. According to a previously developed rating scale, emerging microbiological growth on the probes was assessed. The measurements of the replicates were averaged and graphically evaluated in the course of the biological development on the surfaces. All biocide equipped coatings showed a retarded growth as expected, but characteristic differences could be shown, which could be assigned to different features of the coatings and to different combinations of biocides. Further observations could be linked to elevated tolerances of special microorganisms. Due to known efficiency gaps of different biocides it is already common use to apply adjusted combinations. The important challenge is a careful and sustainable use of

N. Krueger (✉) · W. Hofbauer · M. Krus · C. Fitz · F. Mayer · A. Melzer · K. Breuer
Fraunhofer Institute for Building PhysicsHolzkirchen Branch (IBP), Valley, Germany
e-mail: nicole.krueger@ibp.fraunhofer.de

W. Hofbauer
e-mail: wolfgang.hofbauer@ibp.fraunhofer.de

M. Krus
e-mail: martin.krus@ibp.fraunhofer.de

C. Fitz
e-mail: cornelia.fitz@ibp.fraunhofer.de

F. Mayer
e-mail: florian.mayer@ibp.fraunhofer.de

A. Melzer
e-mail: arnulf.melzer@ibp.fraunhofer.de

K. Breuer
e-mail: klaus.breuerlaus.breuer@ibp.fraunhofer.de

V. P. de Freitas and J. M. P. Q. Delgado (eds.), *Hygrothermal Behavior, Building Pathology and Durability*, Building Pathology and Rehabilitation 1,
DOI: 10.1007/978-3-642-31158-1_3,

biocidal substances with regard to durability, energy consumption and environmental protection.

Keywords Biocides · Biological colonisation · Retarded growth

1 Introduction

Unwanted emergence of microorganisms on building surfaces, especially on the surface of external thermal insulation composite systems (ETICS), is a known problem (Hofbauer et al. 2006, 2009; Fitz et al. 2007; Krus et al. 2007). According to the current literature and the public understanding the influencing variables can be classified in the following groups: global climate change, air pollution, geographical location, exposition of a building, construction details and the choice of materials. Not all of them can be influenced by construction techniques. Present state of the art does not allow guaranteeing the exemption of visually unacceptable biological growth on building coatings under all circumstances and for a period of more than a few years without the use of biocides (Lindner 2005). To be effective against the target organisms an active compound has to be water soluble. Otherwise it could not be incorporated in the living cells of the target organisms and would be ineffective. This water solubility again limits the durability of biocides for the film conservation in coatings for façades that are exposed to rain or condensate. In order to achieve an appropriate and ecological use of biocides in building coatings it is important to know about their durability and effectiveness in combination with different matrices (binders).

In Germany usually it is granted, that a façade shall be free of complaints at least for five years. For Europe the Biocidal Product Directive (BPD), demands to reduce the amount of biocides which may enter the environment as far as possible. Until now, there were only limited data available on the long-term field performance of biocidal coatings. Therefore all parties involved showed big interest in having products tested with regard to performance in a long term investigation.

Biological infestation of building surfaces is an extensive and growing problem. While some materials like metals, for instance, can suffer severe damage caused by biodegradation and biofouling (e.g. von Rège and Sand 1996) growth at façades usually is mainly a visual problem. But if optical degradation of a surface by excessive soiling or microbial growth happens soon after the attachment of a coating this is regarded as damage and may lead to claim of recourse. Due to effective regulations big efforts are demanded for the licensing of new substances. Therefore currently only a limited range of active substances against façade microorganisms is available.

For all these reasons in 2004 an interdisciplinary publicly funded project on effectiveness and durability of biocides in building coatings was started at the Fraunhofer-Institute for Building Physics. In cooperation with partners from industry material components, building coatings and biocide combinations were chosen and probes for outdoor weathering experiments were designed.

Unwanted biological staining of building surfaces may consist of quite different organisms, as there are algae, fungi, lichens and bryophytes (Hofbauer et al. 2003). Bacteria are also involved, but usually they do not form visible growth patterns on modern building surfaces. The durability of biocidal products is exposed to different influences like climate (day course and year course) and material and construction properties. Apart from this also biocide tolerance of microorganisms and biodegradation of biocides form additional important factors. The contribution of bacterial degradation to the degradation of biocides may be significant (Cook and Dickinson 2004). For Iodopropynyl butylcarbamate (IPBC) it was already shown, that some wild isolates of bacteria from treated wood are more effective and more resistant to the compound than other strains (Cook and Dickinson 2004). Fungicidal compounds may have gaps in their activity spectrum. It is already known, that in case of Carbendazim some fungi, like *Alternaria* spp., *Ulocladium* spp., and other, show an increased tolerance (e.g. Bollen et al. 1983; Hassal 1990; Deising et al. 2008).

Because of the biological diversity at infested façades and due to known efficiency gaps of different biocides it is already common to apply adjusted combinations. Aerophytic microorganisms, like façade algae may be capable to follow different survival strategies to avoid or tolerate biocidal ingredients. An example of an algae strain with increased biocide tolerance found will be presented and different features of aerophytic microorganisms and the biological effects against the target organisms of typical biocides are discussed.

2 Materials and Methods

2.1 Biological Assessment of Long Term Exposed Specimen

The investigations presented are part of a four years outdoor study. Special specimens with the structure of an external thermal insulation composite system (ETICS), finished with different coatings (a plaster or a plaster covered with paint) were designed. Five different kinds of plaster and three different types of paints were used, as shown in Table 1.

Four different biocide combinations were added to the surface coatings. Every biocidal mixture consisted of three active ingredients. Two ingredients were always the same for all combinations, a third ingredient was added from a choice of four (see Table 2). The biocide combinations were added in concentrations of 0.5 mass-percent to the plasters and 1.0 mass % to the paints.

Approximately 350 test specimens based on the structure of an ETICS and facing west were exposed to real climatic conditions at two locations in Germany, most of them in Holzkirchen (Fig. 1) in Upper Bavaria north of the Alps. The

Table 1 Tested coatings

Coating	Plasters	Paints
With inorganic binder	Lime cement plaster	
	Silicate plaster	Dispersion type silicate paint
With organic binder	Styrene acrylate based	Styrene acrylate based emulsion paint
	Terpolymer based	
	Silicone resin based	Silicon resin based paint

Table 2 Tested biocide mixtures

Obligate compounds	Variable compounds	Mixture
Terbutryne	Zinc-pyrithione	1
Octylisothiazolinone (OIT)	Iodopropynylbutylcarbamate (IPBC)	2
	Dichloro-Octylisothiazolinone (DCOIT)	3
	Carbendazime	4

Fig. 1 Outdoor exposition of the specimen in Holzkirchen

climate is characterized by three main factors: heavy driving rain (from the west), temperatures below zero for long periods during winter and high daily differences in temperature caused by radiation and occasional warm winds from the south (Foehn). The second location in western Germany, Ernsthofen near Darmstadt, lies in a basin next to a small river and the local climate is somewhat warmer and moister than in Holzkirchen.

Apart from chemical and physical measurements (Breuer et al. 2011, 2012; Schwerd 2009; Schwerd et al. 2011) the development of biological growth on biocidal surfaces in comparison to non biocidal surfaces of the same kind was assessed by different methods. The composition and development of the biological

Table 3 Rating scale for the assessment of surface growth at the specimen according to Hofbauer et al. (2003), simplified

Rating	Occurring microbiological growth resp. microbial growth covering examined area on an average of
10	87.5–100 %
9	75–87 %
8	62.5–75 %
7	50–62,5 %
6	37.5–50 %
5	25–37,5 %
4	5–25 %
3	Less than 5 %
2	Few single spots (less than 5 %)
1	1–3 single spots
0	0 %–no spots

Uncoloured cells indicate minor growth intensity that would not lead to complaints (additionally separated by a red dashed line), light green shaded cells are used at a transition area where complaints may start and intense green coloured cells mark growth intensities that surely would provoke complaints

processes were recorded regularly. A visual characterisation of the pattern of the infection with microorganisms and succession was performed. The assessments took place in a one month/two months (first location) and half year cycle (second location), respectively, under the use of a semi quantitative assessment scale according to Hofbauer et al. (2003), see Table 3. The lower categories of this rating scale provide a minute resolution of the first colonization whereas the upper categories follow the percentage of coverage by growth. A rating between zero and three will be hardly recognised by an untrained person, because there is very little growth. A rating from six to ten is recognizable also by laymen as disfigurement and would cause complaints and rating four and five form a transition level where complaints might occur.

In addition, developing growth was further recorded by a specially developed visual-optical raster scan and documented by photographs. The growth recorded was also categorised if consisting of fungi, algae or both.

More detailed measurements and investigations were performed which will be presented elsewhere. For instance, biological measurements were completed by physiological investigations (efficacy-test) and a complex quantitative and taxonomical analysis of the microorganisms colonizing the different coating-systems. Furthermore also an ecological analysis and a resistance-test accomplished the results of the visual observations.

Table 4 Characterization of the used biocides

Active ingredient (CAS-number)	Substance category	Properties
IPBC (55406-53-6)	Carbamate	Broad spectrum of activity, fungicide, unstable under UV-radiation, high pH-values and presence of catalysts (metals like Co)
Carbendazim (10605-21-7)	Benzimidazole-derivative	Fungicide, high activity (but not against *Alternaria*), chemically stable
OIT (26530-20-1)	Isothiazolinone	Broad spectrum of activity, high water solubility and mobility
DCOIT (64359-81-5)	Isothiazolinone	Broad spectrum of activity, high volatility, unstable, hydrolysis at pH-values >8.5
Zinc-pyrithione (13463-41-7)	Pyridine-derivative	Very broad spectrum of activity, fungicide and algicide, in alkaline systems possibility of complexation with calcium, possibility of discolouration with other metals, hydrolysis at pH-values >9
Terbutryne (886-50-0)	Triazine-derivative	Algicide, high activity, sublimation possible, dependent on the microorganism partly high concentrations necessary, otherwise metabolizable and possibility of formation of resistances

2.2 Biological Mode of Action of Investigated Biocides

To achieve a more complete biocidal effect against the target organisms it is state of the art to employ combinations of different active compounds. In this project the following compounds were used for their known activity against fungi: Carbendazim, IPBC and two different isothiazolinone derivatives, OIT and DCOIT. Zinc-pyrithione functions well against fungi and algae, whereas Terbutryn is active mostly against plants. For a comprehensive overview of the properties of the different biocides see Table 4 (see also Table 2).

Figure 2 illustrates the mode of action of ingredients used against fungi. Isothiazolinones especially block thiol groups of enzymes and therefore hampers special processes of the metabolism of the cell (e.g. Collier et al. 1990, 1991; Nicoletti et al. 1993; Williams 2007). By impairing the development of the nucleus spindle Carbendazim interferes with the cell division (e.g. Davidse 1973, 1986; Davidse and Flach 1977; Hassal 1990; Marshall 2002). The blocking effect of IPBC on proteins and membranes is not fully understood yet, but it is thought to be connected to formation of elementary iodine (Juergensen et al. 2000). Pyridine derivatives present a dysfunctional property on the proton pumps of the whole cell and also at important cell organelles as are the mitochondria and thus blocking the energy supply of the metabolism (Chandler and Segel 1978).

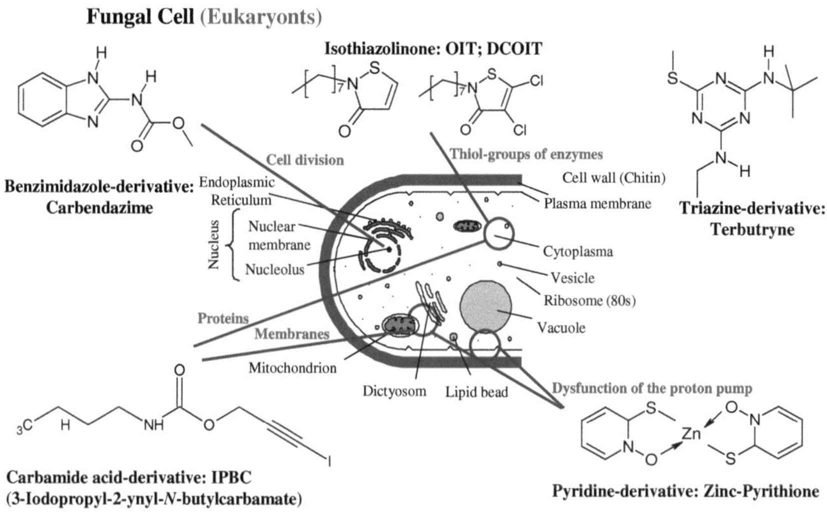

Fig. 2 Mode of action of the used biocides against fungi

Figure 3 presents the mode of action of the used biocidal ingredients against algae. Whereas most other discussed biocides have a pronounced activity against fungi, pyridine derivatives also inhibit plant (algal) growth by interfering also with the proton pumps of the chloroplasts (Chandler and Segel 1978). The triazine derivative Terbutryn is especially added for its activity against plant growth (e.g. Moreland et al. 1959; Shimabukuro 1967; Hassal 1990; Varsano et al. 1992). It directly hinders important steps in photosynthesis and therefore causes gross damage to the photosynthetic organelles because the collected energy from sunlight can not be processed. Terbutryn is a very potent compound, it can also be used to control weeds, but in a different kind of application (e.g. Riemer and Trout 1980; Lindner 2005).

2.3 Investigation of a Possible Biocide Tolerance of Isolated Algae Strains

Apart from chemical and physical measurements [for details see Schwerd (2009)] the development of biological growth at biocidal surfaces in comparison to non biocidal surfaces of the same kind was assessed by different methods. In addition to a permanent control of the visual condition of the west facing surfaces of the specimen according to Hofbauer et al. (2003) and regular microbiological exploration of the surfaces by dilution plating and direct plating according to Hofbauer (2007) also ecophysiological investigations were performed.

The biocide tolerance of unialgal isolates of algae that were recovered from biocidal surfaces was measured in comparison to standard test algae. Batch cultures

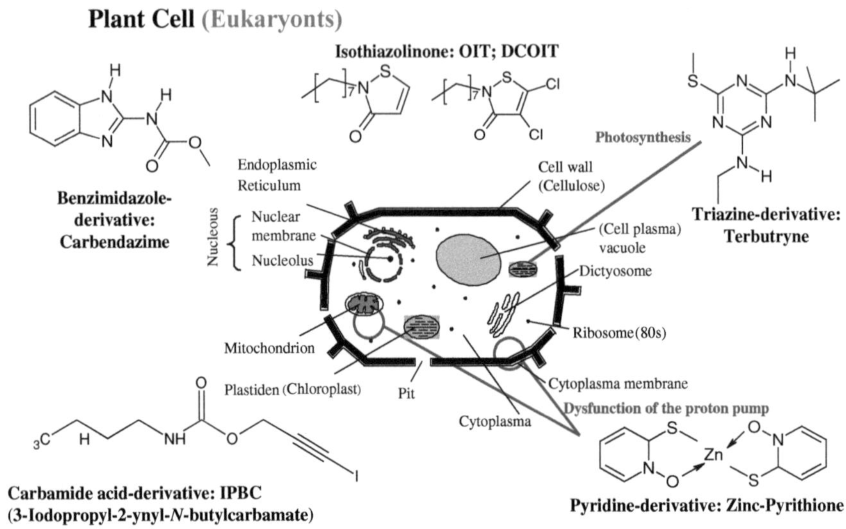

Fig. 3 Mode of action of the used biocides against algae

of new algae isolates from biocidal surfaces and common green algal test strains from the Sammlung von Algenkulturen Göttingen (SAG, Culture Collection of Algae Göttingen) in a mixture, *Stichococcus bacillaris* SAG 379-1a and *Klebsormidium flaccidum* SAG 335-5, were grown in Bold′s Basal medium, modified (Bischoff and Bold 1963; Ettl and Gärtner 1995), and maintained at 16 °C, with light supplied by cool-white fluorescent tubes (Osram) on a 12:12 h light: dark cycle. Room temperature Chlorophyll a (Chl a) fluorescence was measured under vivo conditions at 23 ± 2 °C using a pulse-amplitude-modulated fluorometer (Walz). Non-photochemical Quenching was calculated as described in Cousins et al. (2002).

3 Results

3.1 Biological Assessment of Long Term Exposed Specimen

As shown in previous studies (Hofbauer et al. 2006; Hofbauer 2007), development of growth followed cyclic fluctuations (Fig. 4). In humid times of the year (in Central Europe for example in autumn) a more intense development of growth is recognized whereas in dry and warm or dry and cold times (in Central Europe during mid summer and late winter) a stagnation or even reduction of growth occurs. Once growth is present it usually never disappears completely again. During the next convenient climate phase the development will start again from a higher level. Figure 4 shows a scheme of a typical cyclic fluctuation of growth development with a wavelike increase.

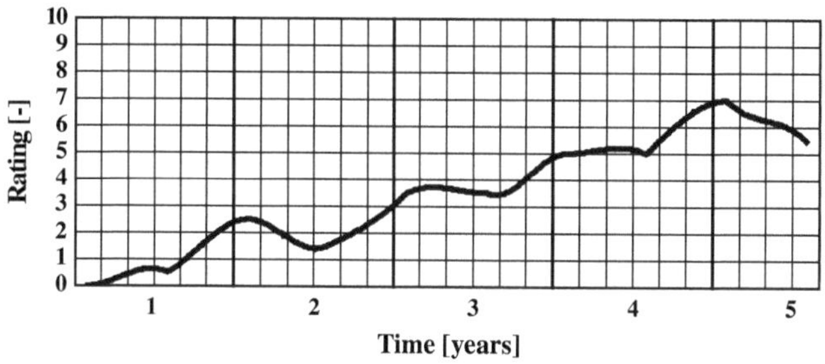

Fig. 4 Typical growth development in phases as measured at a non biocidal surface during the investigations

In general the efficiency of the used biocide-mixtures is confirmed as the growth development on all biocidal equipped coating systems is clearly retarded. At some biocidal variants even after 4 years of exposition the growth was rated inconsiderable or stagnating. On some optical conspicuous variants only fungal growth established, especially of dematiaceous fungi (mould fungi with dark pigmentation) until the end of the project.

A different effectiveness of biocides with regard to different binding agents was observed. Surface growth may occur earlier and more intense in course of time on specimens with comparable high total biocide content in comparison to variants with much lower content at the same time. This effect could be a result of a different withholding capacity for biocides or a different release, respectively.

Specimen with silicon resin based variants turned out to be quite durable against microbiological growth, as shown in Fig. 5. Until the end of the observation time growth intensity on probes of all biocide combinations were rated as inconsiderable, below level 3 according to Table 3. Even at the specimens without biocide the critical rating of 3 was reached not until after 2 years of exposition.

The ratings outlined in Fig. 5 comprise approximate-curves based on geometric mean values from the measurements. The rating of the growth intensity is shown on the ordinate, the numbers of the abscissas show the number of month elapsed. The "critical" zone (the lowest rating at which from very critical observers complaints are possible) is marked by a red dashed line in the diagrams (Table 3, Figs. 5, 6).

Unlike previous variants, specimens with plaster coating based on terpolymer showed already significant growth situations more early both at biocide equipped and non-equipped variants, as shown in Fig. 6. Within the measured period the "critical" rating of three is reached even by all biocide equipped systems.

The lime cement plaster could not be rated conclusively, because the biocidal effect was overlaid by other effects, for instance by chalking and weathering because of heavy driving rain or hail damage.

While in some coating systems all active biocidal components show a similar effect, in some coating systems special biocidal combinations acted better than other. In general, systems with paint coating showed better results than systems

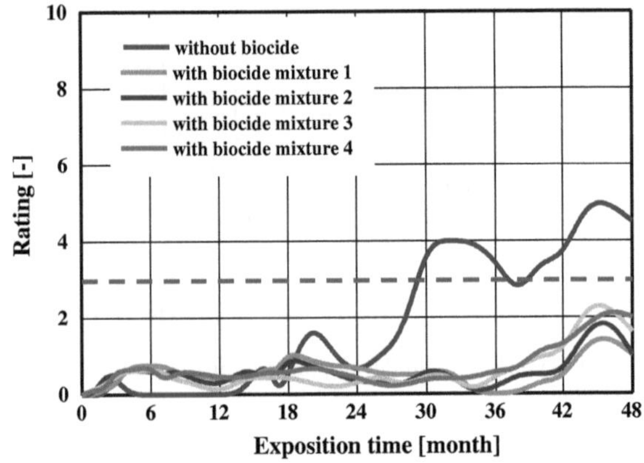

Fig. 5 Observed growth on silicon resin variants (for mixture composition see table 2)

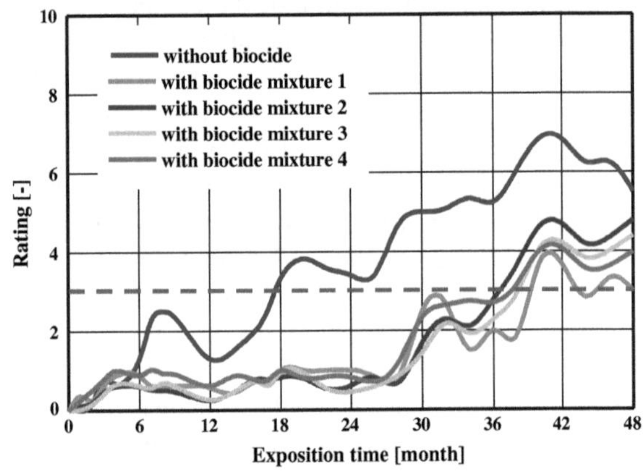

Fig. 6 Observed growth on terpolymer variants (for mixture composition see table 2)

without paint coating. The paint coatings were equipped with twice as high concentration of active compounds than the render. But because of the much lower thickness of the paint coatings the total content of active compounds is less than in renders.

3.2 Investigation of a Possible Biocide Tolerance of Isolated Algae Strains

Continuous visual investigation of the different surfaces revealed a general effectiveness of all tested biocide combinations but different patterns of starting growth processes according to different biocide combinations and different coating systems. Qualitative and quantitative microbiological investigations of the surface of the coatings showed different development at the different coating systems with characteristic groups of organisms. Whereas at the non biocidal variants for

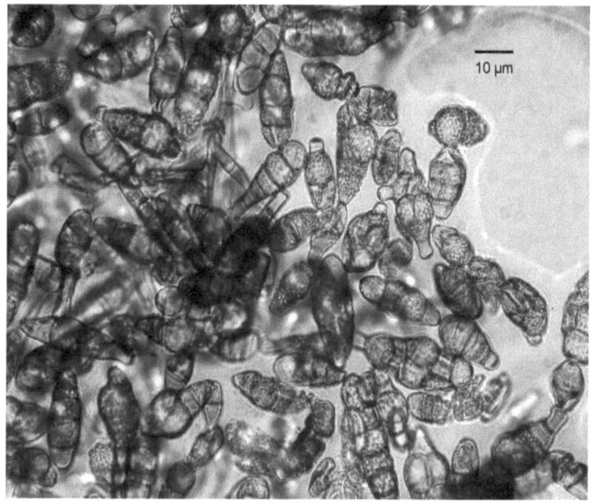

Fig. 7 Microscopic picture of typical muricate spores of *Alternaria alternata*, a fungus known for its relatively high tolerance against the biocide Carbendazim

comparison growth started much earlier and showed a much higher biodiversity, growth at biocidal coatings was rather sparse and showed only limited biodiversity. Selected and peculiar results of these investigations will be highlighted here.

Non biocidal variants show very often growth of the fungal species *Cladosporium* (several species) and *Epicoccum purpurascens*. Also *Alternaria alternata* appears quite frequently, but not dominantly. Apart from the most frequent and dominant species a broad range of different pioneer species can be found. On surfaces of the biocidal variants much fewer species could be isolated. It is conspicuous that on biocidal surfaces that are equipped with a biocide combination with Carbendazim as second fungicide after gross reduction of OIT the fungal genera *Alternaria* (Fig. 7) and *Ulocladium* (Fig. 8) become predominant in the otherwise very sparse surface flora. These fungi are already known for their higher tolerance against Carbendazim. A further quite abundant species at these surfaces is also *Phoma glomerata*, which, according to results presented here, can be assumed as also being quite tolerant to Carbendazim. Other potentially new tolerances of fungi are assumed. Details shall be presented elsewhere.

After a comparable short time and apart from numerous different species of fungi non biocidal variants show also a quite high diversity in algal species. At biocidal surfaces, if there was any algal activity measurable within the course of the project, almost only strains of the green algal genus *Diplosphaera* (Fig. 9) were found. This was interpreted as a potential tolerance of these algal strains against the algicide Terbutryn. Therefore measurements of room temperature chlorophyll a (Chl a) fluorescence under in vivo conditions using a pulse-amplitude-modulated fluorometer were performed. Relative efficiency of non-photochemical quenching (Fig. 10) shows that the new isolates (*Diplosphaera* sp.) are much less affected by the biocidal activity of Terbutryn. This result indicates that these isolates seem to be quite tolerant against Terbutryn, a common algicide used in coatings.

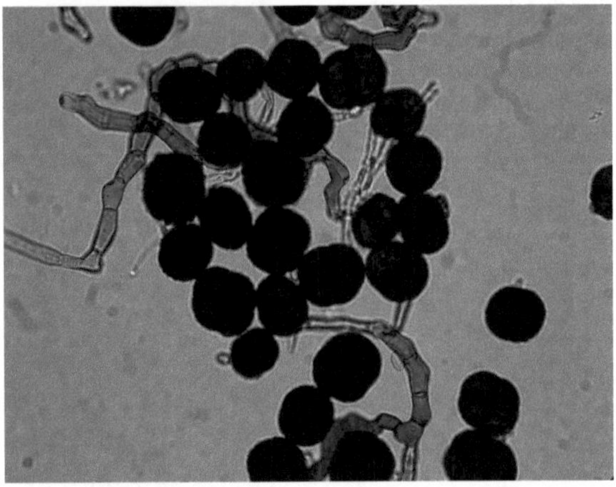

Fig. 8 Microscopic view of *Ulocladium atrum*, another fungus that is quite tolerant against Carbendazim

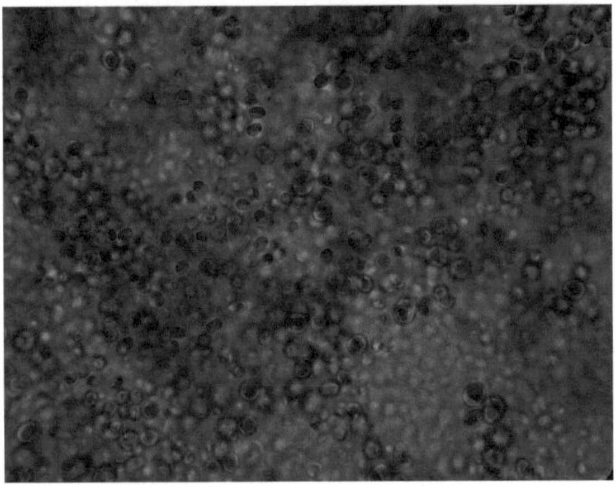

Fig. 9 Isolated strain of *Diplosphaera* sp. (microscopic picture at high magnification), an alga that is supposed to possess an elevated tolerance against Terbutryn

4 Discussion

Results show, that the concentration levels of active compounds in a coating alone do not allow direct assessment of the effectiveness against biological growth. The interaction between the type of biocidal composition, the type of coating (binder) and building physical conditions is of essential importance. Coatings that are generally less colonized by microorganisms because of their chemical constitution are protected by lower total amounts of active compounds. It is important to supply the immediate surface of a coating with effective concentrations of biocides. Therefore coating systems with a pronounced withholding capacity for biocides may lead to an undersupply at the surface.

Based on the chemical character of the biocides used (e.g. Terbutryn, IPBC) the formation of even more active transformation products from the biocidal agents is possible (Breuer et al. 2011). Transformation may be due to effects caused by

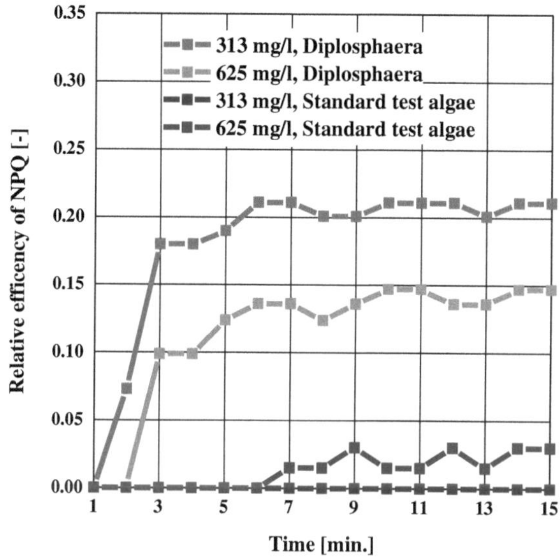

Fig. 10 Comparison of the relative efficiency of the non-photochemical quenching shows that photosynthesis of an isolate of *Diplosphaera* sp. from a biocidal surface is much less affected by different concentrations of Terbutryn than standard test algae

temperature, oxidation, hydrolysation, sunlight, etc. These processes could cause that the concentration of the initially used biocidal compounds may already be decreased but the protection of the façade is still given through the potentially longer lasting biocidal effective transformation products.

Taxonomical and ecological analyses of the species of microorganisms found in the early stages of colonization of the surfaces confirmed that according to the various biocidal compositions and coating systems different ecological groups of organisms are present.

In case of a decrease or loss of an active compound gaps in the action spectrum of the other biocidal compounds became obvious by the appearance of typical species. For example *Alternaria* spp., *Ulocladium* spp., are already known for an increased tolerance against Carbendazim (e.g. Bollen et al. 1983; Hassal 1990; Deising et al. 2008). Therefore specimen equipped with a biocidal combination comprising of Terbutryn, OIT and Carbendazim were earlier colonised by the two fungi mentioned because OIT diminished faster than other biocides.

Results presented show that biocidal equipped paints are very effective in protection of building coatings against unwanted growth of microorganisms. If only the paint finish is equipped with biocides and the underlying render is not the total amount of biocides necessary may be reduced significantly.

While bacteria are hardly recognised for visual disfigurement of modern building surfaces they are well known for developing resistance against various bactericidal compounds. The so called "super bug", a multi resistant pathogen germ is spreading in hospitals all over the world, for instance. For coatings bacteria are mainly a problem in pot conservation and less for film conservation of building coatings.

In Central Europe mainly eukaryotic algae dominate biological stains on modern façades (e.g. Barberousse et al. 2007; Hofbauer 2007). Therefore it is not surprising, that the primary algal colonization that was found in this study comprises almost only (eukaryotic) green algae. But it was a surprise that in early algal colonization of biocidal surfaces only strains of *Diplosphaera* could be found. Since the only biocide against algae that was employed in the project was Terbutryn it seems obvious that the isolated algae strains might have a higher tolerance against this compound. This could be confirmed by first measurements. Further detailed investigations are planned and will be completed by the analysis of further strains of this algal genus from other surface samples.

Terbutryn is a chemical measure against unwanted growth of algae that has a general detrimental effect on photosynthesis and does have impact on all photosynthetic active organisms (e.g. Moreland et al. 1959; Shimabukuro 1967). Previous documented tolerances of algae against algicides were only known from laboratory experiments with artificial produced resistances or from aquatic environments (e.g. Riemer and Trout 1980; Collier et al. 1990; Hassal 1990; Varsano et al. 1992; Lindner 2005). The data shown reveal for the first time that a higher tolerance against algicides can also be developed by aerophytic alga that might colonize building surfaces. Since biocidal action means a high evolutional pressure it is to be assumed that this algal form might spread in the future.

In order to achieve an appropriate and ecological use of biocides in building coatings it is important to know about possible tolerances of the target organisms. Due to known efficiency gaps of different biocides it is already common to apply adjusted combinations. Aerophytic microorganisms, like façade algae may be capable to follow different survival strategies to avoid or tolerate biocidal ingredients.

The above discussed tolerances of algae against biocides are direct resistances on the basis of biochemical changes (mutation) within the cells. But algae and fungi do have some other possibilities to escape the activity of biocides. Many algae produce mucopolysaccharides that cover their cells or filaments. For some organisms it is already assumed that their surrounding mucilage can prevent toxic substances reaching the cells (e.g. Scott et al. 1996). This is also thinkable for biocides, but not yet confirmed and should be a target of further studies.

Another way of reducing effectiveness of biocides is biodegradation. It is possible, even desirable in the sense of ecotoxicological aspects, that biocides may be degraded by bacteria. If algae and fungi live together with bacteria of this special kind in a consortium this may effectively reduce biocidal content in their immediate surrounding.

Many plasters and renders that are used today do contain structure kernels that shall give the finished surface a pleasant structure. Structure kernels usually are made from natural rocks. If these kernels are weathered out at the surface they might be a place where growth could start if conditions otherwise would be favourable.

For the future it will be a great challenge to bring all demands on biocides that are used for building coatings together. An appropriate and ecological use of biocides shall result in a more specific and responsible employment of ecotoxicologic bearable concentrations. Current regulations make efforts for development of new chemicals that are active against the target organisms quite costly. But emerging tolerances of target organisms against biocides currently used demand higher concentrations employed or a change in compounds. It will not be possible to keep building surfaces free from unwanted growth of microorganisms at all for all future by chemical methods, if not completely new approaches are to be developed such as for example encapsulation, nanotechnology, natural antagonists, and quorum sensing.

Acknowledgments The research presented is sponsored by the Bundesministerium für Wirtschaft und Technologie on account of a decision of the Bundestag of Germany, Federal Republic of Germany, Projektträger Jülich and partners from industry.

References

Barberousse, H., Tell, G., Yéprémian, C., Couté, A. (2006): Diversity of algae and cyanobacteria growing on building façades in France. Algological Studies 120: 81–105

Bischoff, W.W., Bold, H.C.: Phycological studies IV. Some soil algae from enchanted rock and related algal species. Univ. Texas Publ. **6318**, 1–95 (1963)

Bollen, G.J., van der Hoeven, E.P., Lamers, J.G., Schoonen, M.P.M.: Effect of benomyl on soil fungi associated with rye. 2. Effect on fungi of culm bases and roots. Neth. J. Pl. Pathol. **89**, 55–66 (1983)

Breuer, K., Hofbauer, W., Krus, M., Scherer, C., Schwerd, R., Krueger, N., Mayer, F., Sedlbauer, K.: Bedeutung des bioziden Wirkstoffeinsatzes bezüglich der Dauerhaftigkeit von Fassadenbeschichtungen. In: Venzmer, H. (ed.) Fassadensanierung—Praxisbeispiele, Produkteigenschaften, Schutzfunktionen, pp. 53–79. Beuth, Berlin (2011)

Breuer, K., Mayer, F., Scherer, C., Schwerd, R., Sedlbauer, K.: , Wirkstoffauswaschung aus hydrophoben Fassadenbeschichtungen: verkapselte versus Unverkapselte Biozidsysteme. Bauphysik **34** (2012 in press)

Chandler, C., Segel, I.: Mechanisms of the antimicrobial action of pyrithione: effects on membrane transport, ATP levels and protein synthesis. Antimicrob. Agents Chemother. **14**(1), 60–68 (1978)

Collier, P.J., Ramsey, A.J., Austin, P., Gilbert, P.: Growth inhibitory and biocidal activity of some isothiazolinone biocides. J. Appl. Bacteriol. **69**, 569–577 (1990)

Collier, P.J., Austin, P., Gilbert, P.: Isothiazolinone biocides: enzyme-inhibiting pro-drugs. Int. J. Pharm. **74**, 195–201 (1991)

Cook, S.R., Dickinson, D.J.: Studies on the toxicity of IPBC and other biocides to bacteria. Proceedings of 35th Annual Meeting Ljubljana, Slovenia 6–10 June 2004. The International Research Group On Woos Preservation, Section 1 Biology, IRG/WP 04-10544, pp. 1–20 (2004)

Cousins, A.B., Adam, N.R., Wall, G.W., Kimball, B.A., Pinter JR, P.J., Ottman, M.J., Leavitt, S.W., Webber, A.N.:, Photosystem II energy use, non-photochemical quenching and the xanthophyll cycle in *Sorghum bicolor* grown under drought and free-air CO_2 enrichment (FACE) conditions. PLANT CELL ENVIRON. **25**, 1551–1559 (2002)

Davidse, L.C.: Antimitotic activity of methyl benzimidazol-2-yl carbamate (MBC) in *Aspergillus nidulans*. Pestic. Biochem. Physiol. **3**, 317–325 (1973)

Davidse, L.C.: Benzimidazole fungicides: mechanism of action and biological impact. Ann. Rev. Phytopathol. **24**, 43–65 (1986)

Davidse, L.C., Flach, W.: Differential binding of Benzimidazol-2-yl Carbamate to fungal tubulins as a mechanism of resistance to this antimitotic agent in mutant strains of *Aspergillus nidulans*. J. Cell Biol. **72**, 174–193 (1977)

Deising, H.B., Reimann, S., Pascholati, S.F.: Mechanism and significance of fungicide resistance. Braz. J. Microbiol **39**, 286–295 (2008)

Ettl, H., Gärtner, G.: Syllabus der Boden- Luft- und Flechtenalgen. Gustav Fischer, Stuttgart (1995)

Fitz, C., Krus, M., Sedlbauer, K., Hofbauer, W., Breuer, K.: Mikrobieller Bewuchs auf Fassaden—Veralgungen lassen sich erklären und vermeiden. Deutsche BauZeitschrift **9**, 84–88 (2007)

Hassal, K.A.: The Biochemistry and uses of Pesticides. Structure, metabolism, Mode of Action and Uses in Crop Protection. Second Edition. VCH, Weinheim, New York (1990)

Hofbauer, W.: Aerophytische Organismen an Bauteiloberflächen, Ph.D. Thesis, Innsbruck University, Austria (2007)

Hofbauer, W., Breuer, K., Sedlbauer, K.: Algen, flechten, moose und farne auf fassaden. Bauphysik **25**(6), 383–396 (2003)

Hofbauer, W., Fitz, C., Krus, M., Sedlbauer, K., Breuer, K.: Prognoseverfahren zum biologischen Befall durch Algen, Pilze und Flechten an Bauteiloberflächen auf der Basis bauphysikalischer und mikrobieller Untersuchungen, Bauforschung für die Praxis, vol. 77, Fraunhofer IRB Verlag, Germany (2006)

Hofbauer, W., Breuer, K., Sedlbauer, K. Quelle für Lebensvorgänge. Algen, Flechten, Moose und Farne auf Fassaden. Der Maler und Lackierermeister **2**, 14–19 (2007)

Juergensen, L., Busnarda, J., Caux, P.-Y., Kent, R.: Fate, behaviour, and aquatic toxicity of the fungicide IPBC in the Canadian environment. Environ. Toxicol. **15**, 201–213 (2000)

Krus, M., Hofbauer, W., Lengsfeld, K.: Mikrobielles Wachstum auf WDVS als Folge der neuen Bautechnik? Der Bausachverständige **3**(2007), 27–30 (2007)

Lindner, W.:'5.14 Surface coatings. In: Paulus, W. (ed.) Directory of Microbicides for the Protection of Materials. pp. 347–376, Springer, New York (2005)

Marshall, H.: Carbendazim—Fact sheet. Pestic. News **57**, 20–21 (2002)

Moreland, D.E., Gentner, W.A., Hilton, J.L., Hill, K.L.: Studies on the mechanism of herbicidal action of 2-Chloro-4, 6-bis(etylamino)-s-triazine. Plant Physiol. **34**, 432–435 (1959)

Nicoletti, G., Boghossian, V., Gurevitch, F., Borland, R., Morgenroth, P.: The antimicrobial activity of chlorhexidime, a mixture of isothiazolinones ("Kathon" CG) and cetyl trimethyl ammonium bromide (CTAB). J. Hosp. Infect. **23**, 87–111 (1993)

von Rège, H., Sand, W.: Mikrobielle Werkstoffzerstörung—Biofilm und Biofouling. Simulation mikrobiell beeinflußter Metallkorrosion (MIC) zur Prüfung von Gegenmaßnahmen. Mater. Corros. **47**, 486–494 (1996)

Riemer, D.N., Trout, J.R.: Effects of low concentration of terbutryn on myriophyllum and cabomba. J. Aquat. Plant Manage. **18**, 6–9 (1980)

Schwerd, R.: Verweilverhalten biozider Wirkstoffe in Außenbeschichtungen im mehrjährigen Freilandversuch. Forschungsergebnisse aus der Bauphysik vol. 8, pp. 1–219, Fraunhofer Verlag, Stuttgart (2009)

Schwerd, R., Scherer, C., Mayer, F., Breuer, K.: Biozide in Bautenbeschichtungen—chemische Untersuchungen zur Dauerhaftigkeit. Der Bausachverständige **7**(3), 30–34 (2011)

Scott, C., Fletcher, R.L., Bremer, G.B.: Observations on the mechanism of attachment of some marine fouling blue-green algae. Biofouling **10**, 161–173 (1996)

Shimabukuro, R.H.: Atrazine metabolism and herbicidal selectivity. Plant Physiol. **42**, 1269–1276 (1967)

Varsano, R., Rabinowitch, H.D., Rubun, B.: Mode of action of piperonyl butoxide as herbicide synergist of atrazine and terbutryn in maize. Pestic. Biochem. Physiol. **44**, 174–192 (1992)

Williams, T.M.: The mechanism of action of isothiazolinone biocides. Power Plant Chem. **9**, 14–22 (2007)

Hygrothermal Conditions and Mould Growth Potential in Cold Attics: Impact of Weather, Building System and Construction Design Characteristics

Carl Eric Hagentoft and Angela Sasic Kalagasidis

Abstract Field surveys and experiences of cold attics show that these are risk constructions. Both old and new constructions are facing problems. For old constructions there is a clear correlation with improved attic floor insulations. In this chapter the hygrothermal condition of the attic is simulated both using simplified as well as more advanced models. These models are also used to generate probabilistic assessment of mould growth risk. Validation of the advanced HAM-model using field measurements is also demonstrated.

Keywords Hygrothermal · Moisture · Attic · Mould growth

1 Introduction

In the Swedish building stock 88 % of the buildings have a roof construction with a cold attic and mould is visible in about 15 % of them. These numbers are based on in situ inspections performed in a recent national investigation (Åberg 2011). Mould is present mainly in older single-family houses, but also in attics in new houses (see Fig. 1). Also, an overall increasing trend in mould appearance has been noted.

C. E. Hagentoft · A. S. Kalagasidis (✉)
Division of Building Technology, Department of Civil Engineering, Chalmers University of Technology, 412 96 Gothenburg, Sweden
e-mail: angela.sasic@chalmers.se

C. E. Hagentoft
e-mail: carl-eric.hagentoft@chalmers.se

V. P. de Freitas and J. M. P. Q. Delgado (eds.), *Hygrothermal Behavior, Building Pathology and Durability*, Building Pathology and Rehabilitation 1,
DOI: 10.1007/978-3-642-31158-1_4,

Fig. 1 Example of moisture damaged attics

There are differences in the number of cases per region. Mild and humid climate in the west coastal region seems to be more favourable for mould growth than the cold continental climate in other parts of the country. In the larger region around the city of Gothenburg, which is located at the west coast, mould has been detected in about 60–80 % of samples taken from the attics in single-family houses (Ahrnens and Borglund 2007). Based on the same investigation, there is also a certain correlation between the occurrence of mould in attics and the following working conditions. Thus there is an increased risk for moisture problems in the attic related to:

- Naturally ventilated buildings and mechanical ventilation systems that creates an indoor positive pressure
- Heating system, not using combustion
- Large moisture production indoors (i.e. large moisture supply)
- Roof underlay consisting of wooden board such as plywood—instead of wood
- Increased thickness of the attic ceiling insulation
- Air leakage through the attic floor
- Ventilation of attics in cold and maritime climates.

The high humidity levels are to a large extent a consequence of the increasing demand on energy efficiency. Houses are frequently retrofitted with additional attic insulation, which leads to a colder attic space and hence a higher humidity (Hagentoft et al. 2008). Leaks of indoor air up to the attic through the attic floor, and the under cooling of the roof due to sky radiation, increase the problem (Holm and Lengsfeld 2006; Sanders 2006; Essah et al. 2009). The moist air might condensate at the underlay and small droplets of liquid water can build up. The water will then be absorbed and accumulated in the surface area. High moisture content can even lead to rot.

Another important moisture source influencing the attic hygrothermal condition is the water vapour in the surrounding outdoor air. The advice given to the building sector in Sweden today is to have a not too high or not too low ventilation rate, by outdoor air, of the attic. A too high ventilation rate, in combination with under cooling, results in high relative humidity (Sasic Kalagasidis 2004). Too low ventilation is also risky in case of construction damp or leaky attic floor (Arfvidsson and Harderup 2005, Sanders 2006, Essah et al. 2009). The optimal air exchange rate varies with the outdoor climate, and fixed ventilation through open eaves and/or gable and ridge vents are not always the best choice (Hagentoft et al. 2008, 2010). Alternative ventilation strategies as well as the use of adaptive ventilation is analysed later in this chapter.

We can expect more and more problems in cold attics, as a result of demands for increased energy efficiency and future climate change. It is obvious that the situation today, where the potential for mould damage is inadvertently designed into the structure, is totally unacceptable and that something must be done!

In the next sections, both simplified and advanced simulation models for the hygrothermal condition of attics are presented and compared. A field measurement of two types of cold attics, with controlled and traditional ventilation, is used to validate the advanced simulation model. The deterministic models are used together with Monte Carlo simulations in order to find the probability of high and critical moisture conditions. In the risk assessment, state of the art mould growth models based on the transient hygrothermal condition are used.

2 Probabilistic Analysis of the Attic Using Simplified HAM-Model

A probabilistic model of attic performance is presented in this section using dynamic models for the heat and moisture transfer. Simplified transient models will be used, but in principle, more advanced models can of course be used as well, se later sections. The yearly variations in the attic will be calculated based on measured full year weather data. The Monte Carlo method will be used to calculate the probability distribution for the mould growth potential. One of 30 weather data years (1961–1990) with hourly values will be selected randomly for each simulation. The air exchange rates (1/h) for the attic ventilation by exterior air, n_e, and the fraction of the indoor ventilation rate that flows up to the attic, f_{ia}, will have a fixed value for each yearly simulation. Normally distributed random numbers are assumed.

$$n_e = N(\mu_{ne}, \sigma_{ne})\; f_{ia} = N(\mu_{fia}, \sigma_{fia})\; n_e \geq 0.1\; f_{ia} \geq 0 \tag{1}$$

The indoor moisture supply (kg/m^3) will vary randomly each hour according to:

$$\Delta v = N(\mu_{\Delta v}, \sigma_{\Delta v}) \tag{2}$$

These random numbers are limited to the minimum physically realistic values.

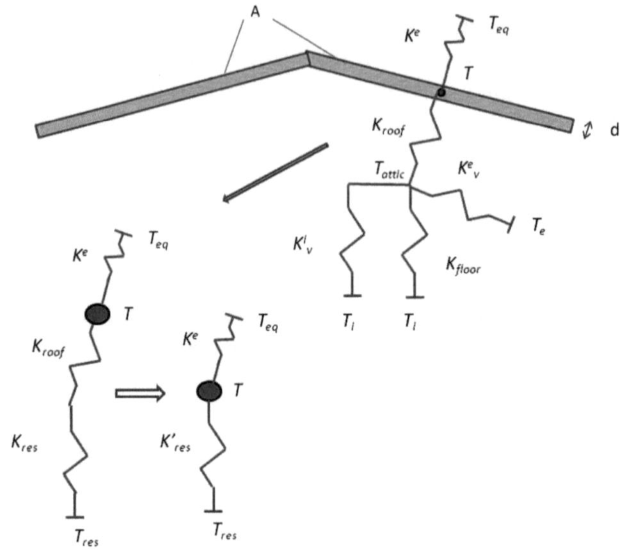

Fig. 2 Thermal network of attic using thermal conductances K (with the unit W/K) and a node representing the heat capacity (J/K) of the wooden roof underlay. The second and third network (*down left*) shows reduced networks

2.1 Thermal Model

In the modelled attic, the roof underlay consists of 0.015 m thick wooden boards with a water and moisture tight membrane facing the exterior (see Fig. 2). The attic floor is thermally insulated with varying air tightness towards the living space underneath.

The attic is ventilated through openings at the gables. The model assumes a common resultant attic air temperature. The heat capacity is located in the roof underlay, i.e. the remaining heat capacity of materials in the attic is neglected. The attic temperature is determined by the heat exchange with the outdoor air through ventilation and with the indoor air through air leakages, as well as by the heat transmission through the attic floor. The temperature, T, of the underlay is determined by the heat exchange with the attic air temperature, the exterior equivalent temperature and the heat storage by the material itself. In this simple model, differences due to the two different oriented sides of the roof are neglected. Basically this corresponds to a low sloped roof.

The indoor temperature is T_i, the exterior temperature T_e and the equivalent temperature for the roof is T_{eq}.

Using thermal conductances K (W/K), the following ordinary differential equation can be found for the wooden roof underlay temperature:

$$Ad\rho c\frac{dT}{dt} = (T_{res} - T)\cdot K'_{res} + (T_{eq} - T)\cdot K^e \tag{3}$$

Here

$$T_{res} = \frac{K_v^e\cdot T_e + K_{floor}\cdot T_i + K_v^i\cdot T_i}{K_{res}} \qquad K_{res} = K_v^e + K_{floor} + K_v^i \tag{4}$$

Here, d(m) represents the thickness of the wooden roof underlay layer, ρc (J/m^{3}K) the volumetric heat capacity of wood, and A (m^2) is the total roof surface area. Reduction of the two serial coupled conductances gives (Hagentoft 2001):

$$1/K'_{res} = 1/K_{res} + 1/K_{roof} \tag{5}$$

The conductances are given by the following expressions:

$$\begin{aligned} &K_{floor} = A_{floor}\cdot U_{floor}\\ &K_v^e = \rho_a c_{pa}\cdot R_a^e \qquad K_v^i = \rho_a c_{pa}\cdot R_a^i\\ &K^e = \frac{A}{1/(\alpha_c+\alpha_r) + d/2/\lambda_{roof}} = \frac{A}{R_{se} + d/2/\lambda_{roof}} K_{roof} = \frac{A}{R_{si} + d/2/\lambda_{roof}} \end{aligned} \tag{6}$$

Here, R_a refers to the ventilation flow rate (m^3/s) and the indices refer to if it is air from the exterior, e, or the interior, i (air leakage from the room below). The volumetric heat capacity of air, at constant pressure, is denoted $\rho_a c_{pa}$ (J/m^{3}K). The thermal conductivity of wood is denoted λ_{roof} (W/mK). The heat transfer coefficients for long wave radiation and convection are denoted α_r and α_c (W/m^{2}K) respectively. The total surface resistances at the interior and exterior side of the roof are denoted R_{si} and R_{se} (m^{2}K/W) respectively.

Expressed as air exchange rates per hour (1/h) we can write the air flow rates through the attic as:

$$R_a^e = V_{attic}\cdot\frac{n_e}{3600} \qquad R_a^i = V_{room}\cdot\frac{n_i}{3600}\cdot f_{ia} \tag{7}$$

The latter express the indoor air leakage to the room as a fraction, f_{ia}, of the room air ventilation rate.

Rewriting (3) we get:

$$\begin{aligned} &\frac{dT}{dt} = -T_{factor}\cdot T + T_{e,factor}\cdot T_e + T_{eq,factor}\cdot T_{eq} + T_{i,factor}\cdot T_i\\ &T_{factor} = \frac{1}{A_{floor}d\rho c}\left(K'_{res} + K^e\right) \quad T_{e,factor} = \frac{1}{A_{floor}d\rho c}K_v^e\frac{K'_{res}}{K_{res}}\\ &T_{eq,factor} = \frac{K^e}{A_{floor}d\rho c} \quad T_{i,factor} = \frac{1}{A_{floor}d\rho c}\left(K_{floor} + K_v^i\right)\cdot\frac{K'_{res}}{K_{res}} \end{aligned} \tag{8}$$

This general way of writing the differential equation will be handy when using our mathematical solver.

The attic air temperature is given by:

$$T_{attic} = \frac{K_{roof} \cdot T + K_{res} \cdot T_{res}}{K_{roof} + K_{res}} \tag{9}$$

The underlay surface temperature is:

$$T_{surf} = T + (T_{res} - T)\frac{K'_{res}}{A_{roof}}\frac{d}{2\lambda} \tag{10}$$

The external equivalent temperature is determined by the exterior air temperature, T_e, the global solar radiation, I_g (W/m^2) and the long wave radiation, I_{lw} (W/m^2).

$$T^{eq} = T_e + R_{se} \cdot \left(I_g \cdot \alpha_{sol} + \alpha_r \cdot (T^r - T_e)\right) \quad T^r = \sqrt[4]{\frac{I_{lw}}{\sigma}} - 273.15(^\circ\mathrm{C}) \tag{11}$$

Here, T^r represents the apparent sky temperature, and σ is the Stefan-Boltzmann constant. The solar absorptance of the roof surface is denoted α_{sol}.

2.2 Moisture Model

The moisture model assumes that the whole moisture storage capacity of the roof is within the roof underlay material and that the roof surface is perfectly moisture tight (see Fig. 3). The diffusion of moisture through the attic floor is neglected. The only way for moisture to come into the attic is by ventilation and through air leakages.

The mass balance of the underlay, assuming no condensation at the roof underlay surface, gives:

$$\begin{aligned} \frac{dM}{dt} &= K^v_{res}(v_{res} - v) = K^v_{res}\left(\frac{v_i \cdot R^i_a + v_e \cdot R^e_a}{R^i_a + R^e_a} - v\right) \\ &= K^v_{res} \cdot \frac{v_i \cdot R^i_a}{R^i_a + R^e_a} + K^v_{res} \cdot \frac{v_e \cdot R^e_a}{R^i_a + R^e_a} - K^v_{res} \cdot v \end{aligned} \tag{12}$$

Here, M (kg), is the total moisture weight of the roof underlay and, v (kg/m^3), is its humidity by volume.

$$\begin{aligned} & R_{a,res} = R^i_a + R^e_a \\ & v_{res} = \frac{v_i \cdot R^i_a + v_e \cdot R^e_a}{R^i_a + R^e_a} \quad 1/K^v_{res} = \frac{1}{R_{a,res}} + \frac{1}{A/Z} \; Z = Z_{si} + \frac{d}{2\delta_v} \end{aligned} \tag{13}$$

Here, Z (s/m), is the vapour resistance between the air and the centre of the underlay layer and δ_v (m^2/s) is the vapour permeability of wood. The interior and exterior humidity by volume are denoted v_i and v_e (kg/m^3) respectively.

For the case of surface condensation we get:

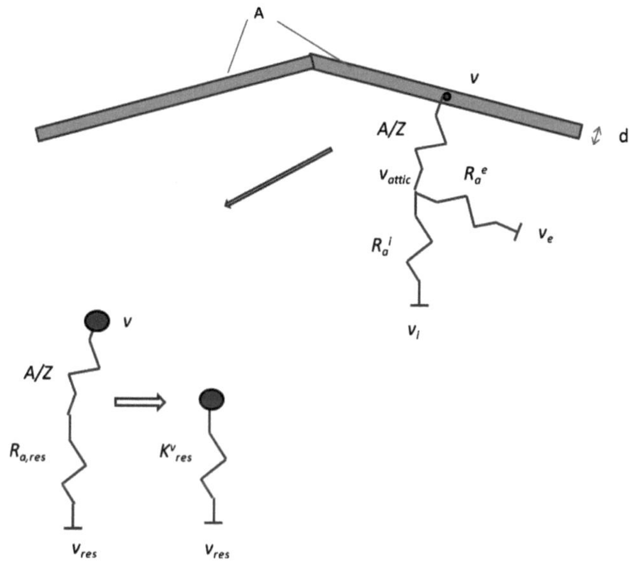

Fig. 3 Moisture network of attic using vapour transfer conductances (with the unit m^3/s) and a node representing the moisture capacity of the wooden roof underlay. The second and third ones show reduced networks

$$\begin{aligned}
&\frac{dM}{dt} = \overline{K}_{res}^{v} \cdot \left(v_{res} - v_{sat}\left(T_{surf}\right)\right) \qquad v_{res} - v_{sat}\left(T_{surf}\right) \geq 0 \\
&\frac{dM}{dt} = \overline{K}_{res}^{v} \cdot \frac{v_i \cdot R_a^i}{R_a^i + R_a^e} + \overline{K}_{res}^{v} \cdot \frac{v_e \cdot R_a^e}{R_a^i + R_a^e} - \overline{K}_{res}^{v} \cdot v_{sat}\left(T_{surf}\right) \\
&1/\overline{K}_{res}^{v} = \frac{1}{R_{a,res}} + \frac{1}{A/Z_{si}}
\end{aligned} \tag{14}$$

Here, v_{sat}, denotes the humidity at saturation. We assume that the condensed water on the surface of the underlay is absorbed by the layer. Assuming a known slope, $\xi(\varphi)$ (–), of the sorption isotherm $w(\varphi)$ (kg/m^3) we can write:

$$\frac{dM}{dt} = A_{roof} d\xi \frac{d\varphi}{dt} \qquad \xi = \frac{dw}{d\varphi} \tag{15}$$

Combining it with (12) and using the definition of relative humidity we have for the case with no condensation:

$$\frac{d\varphi}{dt} = \left(-\varphi_{factor} \cdot v_{sat}(T) \cdot \varphi + v_{i,factor} \cdot v_i + v_{e,factor} \cdot v_e\right)\frac{1}{\xi} \qquad v_{res} - v_{sat}(T_{surf}) < 0$$

$$\varphi_{factor} = K^v_{res}\frac{1}{A_{roof}d}$$

$$v_{i,factor} = K^v_{res}\frac{R^i_a}{R^i_a + R^e_a}\frac{1}{A_{roof}d} \qquad v_{e,factor} = K^v_{res}\frac{R^e_a}{R^i_a + R^e_a}\frac{1}{A_{roof}d} \tag{16}$$

and with condensation:

$$\frac{d\varphi}{dt} = \left(-\overline{\varphi}_{factor} \cdot v_{sat}(T_{surf}) + \overline{v}_{i,factor} \cdot v_i + \overline{v}_{e,factor} \cdot v_e\right)\frac{1}{\xi} \qquad v_{res} - v_{sat}(T_{surf}) \geq 0$$

$$\overline{\varphi}_{factor} = \overline{K}^v_{res}\frac{1}{A_{roof}d} = \frac{\overline{K}^v_{res}}{K^v_{res}} \cdot \varphi_{factor}$$

$$\overline{v}_{i,factor} = \frac{\overline{K}^v_{res}}{K^v_{res}} \cdot v_{i,factor} \qquad \overline{v}_{e,factor} = \frac{\overline{K}^v_{res}}{K^v_{res}} \cdot v_{e,factor} \tag{17}$$

The exterior humidity by volume, v_e, is obtained from weather data. The interior one comes from:

$$v_i = v_e + \Delta v \tag{18}$$

Here, the second term represents the moisture supply, which will be a random variable, see (2).

2.3 Mould Growth Model

A fundamental uncertainty lies in how to evaluate the calculated temperature and relative humidity. What we really want to do is to estimate the probability or risk for mould growth in the attic. There are several studies on this subject. However, there is no standardized or widely accepted method for the evaluation of the mould growth risk. Here, we will base the risk assessment on the method developed by (Hukka andViitanen 1999). As a measure of the mould risk, a mould growth potential m has been used in this paper, which is simply the relative humidity divided by the critical relative humidity for mould growth to start according to the reference:

$$m = \frac{RH}{RH_{crit}}$$
$$RH_{crit} = \begin{cases} 100\ \%, T \leq 0\ °\text{C} \\ -0.00267T^3 + 0.160T^2 - 3.13T + 100.0,\ 0\ °\text{C} \leq T \leq 20\ °\text{C} \\ 80\ \%, T > 20\ °\text{C} \end{cases} \quad (19)$$

Theoretically, mould growth is possible only when $m > 1$. Hence, the development over time of m represents an indicator of risk for the development of mould. By using this parameter different attic designs can be compared.

2.4 Results of Monte Carlo Simulations

The "ode45" Matlab solver for coupled ordinary differential Eqs. (8), (16) and (17) has been used to calculate the hourly value for a whole year for the attic temperature and relative humidity. By combining this with the model for mould growth potential the probability density function can be calculated.

The following data have been used in the simulation:

Weather: Gothenburg, Sweden, $A = 100$ m^2, $d = 0.015$ m, $V_{attic} = 100$ m^3, $V_{room} = 250$ m^3, $n_i = 0.5$ 1/h, $\delta_v = 2{\cdot}10^{-6}$ m^2/s, $R_{si} = 0.13$ m^{2}K/W, $R_{se} = 0.04$ m^{2}K/W, $Z_{si} = 360$ s/m, $\alpha_r = 4$ W/m^{2}K, $\alpha_{sol} = 0.7$, $T_i = 20$ °C, $\varphi_{start} = 0.8$, $\lambda_{roof} = 0.14$ W/mK, $\rho c = 750\,000$ J/m^{3}K.

The sorption isotherm for the wood (IEA, Annex 14 1992) is given by:

$$w(\varphi) = \frac{\varphi}{-0.027\varphi^2 + 0{,}027\varphi + 0.0089} \quad (20)$$

The indoor moisture supply is the same for all scenarios:

$$\Delta v = N(0.004, 0.002)\ (\text{kg/m}^3) \quad (21)$$

For well and less ventilated attics we have the alternatives:

$$n_e = N(7,5)\ n_e = N(2,1)\ (1/\text{h}) \quad (22)$$

For tight, less tight and leaky attic floor we have:

$$f_{ia} = 0\ f_{ia} = N(0.1, 0.1)\ \ f_{ia} = N(0.2, 0.1)\ (-) \quad (23)$$

The attic floor insulation is thick, 0.50 m, or thin, 0.05 m, giving:

$$U_{floor} = 0.08\ \text{W/m}^2\text{K}\ U_{floor} = 0.8\ \text{W/m}^2\text{K}$$

The air exchange rates are constant for each yearly simulation. The indoor moisture supply for each hour of the year is calculated before the actual dynamic

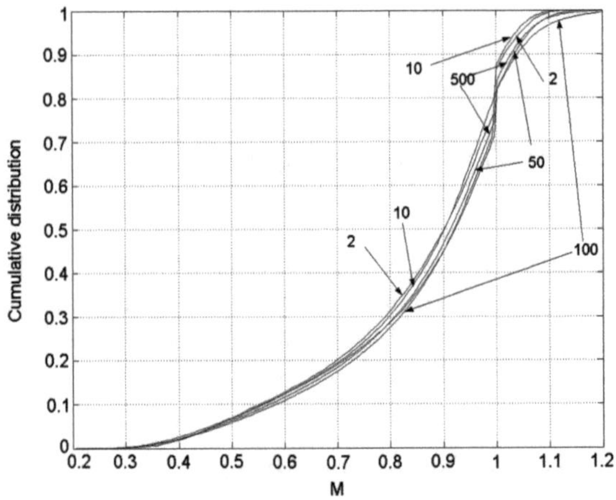

Fig. 4 The cumulative distribution for the mould growth potential of the underlay wooden surface. The scenario with an attic floor having thick insulation is less air tight and where the attic is less ventilated is simulated. The number of samples, i.e. yearly simulations, varies from 2 to 500

Table 1 The probability of exceeding a mould growth potential of one for the underlay wooden surface

Number of yearly samples	2	10	50	100	500
Probability (%)	13.5	13.9	17	18.5	19.6

The scenario with an attic floor having thick insulation is less air tight and where the attic is less ventilated is simulated. The number of samples, i.e. yearly simulations, varies from 2 to 500

simulation for a year starts. The calculation time is approximately 25 s per sample i.e. one year's simulation, on a PC (Intel Core 2 Duo CPU, P9400 @ 2.40 GHz).

Figure 4 shows the cumulative distribution for different samples i.e. number of yearly simulations. The scenario with an attic floor having thick insulation is less airtight and where the attic is less ventilated the scenario is simulated. The results for the probability of exceeding the mould growth potential of one is given in Table 1 as a function of number of yearly samples.

Tables 2 and 3 show the probability of exceeding the mould growth potential of one for the specified scenarios using 50 yearly samples.

3 Hygrothermal Analysis of the Attic Using Advanced HAM-Model

As indicated above, attic ventilation and air infiltration from the dwelling through the attic floor are one of the most influencing parameters on the hygrothermal performance of the attic. The respective air flow rates vary largely in time and mainly due to wind induced pressures differences over the building. Moreover,

Table 2 The probability of exceeding a mould growth potential of one for the underlay wooden surface

	Leaky attic floor (%)	Less tight attic floor (%)	Tight attic floor (%)
$d_i = 0.05$ m	14	12	6
$d_i = 0.5$ m	23	20	13

The scenario for the attic floor insulation, the attic floor tightness varies. The attic is *well ventilated*

Table 3 The probability of exceeding a mould growth potential of one for the underlay wooden surface

	Leaky attic floor (%)	Less tight attic floor (%)	Tight attic floor (%)
$d_i = 0.05$ m	13	11	2
$d_i = 0.5$ m	26	17	9

The scenario for the attic floor insulation, the attic floor tightness varies. The attic is *less ventilated*

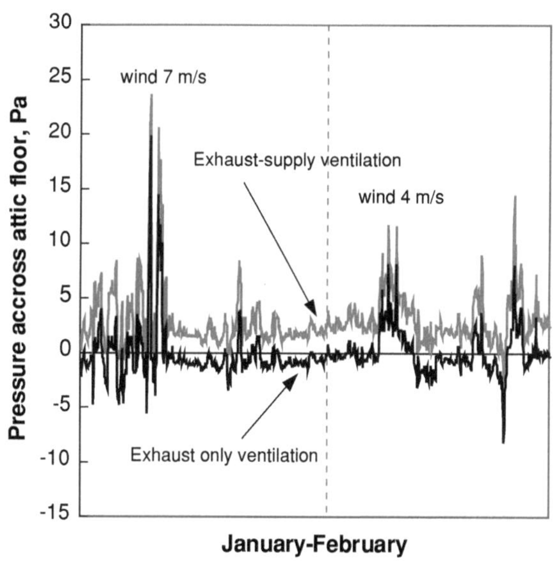

Fig. 5 Pressure differences across the attic floor in a two-storey single family house in respect to the ventilation system in the house. In both cases the attic is ventilated naturally. Positive values denote air inflow from the house to the attic. The results are obtained by the advanced HAM-model of the attic, which is described hereafter

resulting air pressure difference between the attic and the dwelling depends also on air tightness of the house and type of ventilation system in the house (natural, mechanical), as it is shown in Fig. 5. When air pressure in the house is higher than in the attic, air from the house infiltrates to the attic, through the leakages in the attic floor. The opposite is also possible, in which case the air from the attic is sucked into the house. These phenomena could be captured by an advanced HAM-model of an attic, which is presented in this section.

Besides detail modelling of air pressure distribution in the house and calculation of air flow rates through the attic, the advanced HAM-model allows also more freedom in modelling of the attic construction. Slope of the roof sides, materials in the roof and floor, size and orientation of the attic/house, moisture production inside the house are examples of parameters which can be varied freely in the advanced model. Unlike from the simple model, each of the capacity layers is discretized with a number of capacity (discretization) nodes and in the normal direction to the surface (1D approach).

3.1 Advanced HAM-Model of the Attic

For advanced attic simulations a whole building Heat, Air and Moisture (HAM) modelling tool, named HAM-Tools is used. The simulation tool is developed in a Matlab modelling environment using the Simulink toolbox (www.matworks.com). It is developed as a library of block diagrams where each block represents a model of a building component (wall, window), ventilated zone or HVAC system. The models can be combined into more complex building structures, systems and whole buildings by using a set of communication signals, which are specified in the guidelines of the International Building Physics Toolbox (IBPT). HAM-Tools library is developed for research purposes and is publicly available through the Internet (www.ibpt.org).

Since the introduction of this tool, extensive work has been done to validate its use. The accuracy of energy and mass balance models for building components and whole buildings was tested and validated within different research projects (Hagentoft et. al 2004, Woloszyn and Rode 2008). The validation was made by comparing the modelling results with corresponding results from another building simulation programmes and measured data. The experimental results of Samuelson (1995) on ventilation of cold attics were used for the validation of the HAM-Tools model of an attic without consideration in air infiltration from the rooms below the attic (Sasic Kalagasidis 2004b). The calculated hygrothermal states in the attic were in a good agreement with the measured data. The present version of the attic model, which takes into account the pressure inside the house for the calculation of air ventilation and infiltration in the attic, was validated against the experimental results of Hagentoft et al. (2008) on controlled (adaptive) ventilation of cold attics. Details of these validation tests are presented in Sect. 4.

3.1.1 Heat and Moisture Balance in the Attic

The numerical model of the attic is built from the HAM-Tools blocks for building construction. These models for roofs, gables and attic floor are basically one dimensional and support coupled heat, air and moisture balance through multi-layered structures. Temperatures T (°C) and moisture contents w (kg/m^3) in a wall are found from the energy and mass balance equations:

$$c\rho\frac{dT}{dt} = -\frac{\partial q}{\partial x} \text{ and } \frac{dw}{dt} = -\frac{\partial g}{\partial x} \tag{24}$$

where c (J/kg/K) is the specific heat capacity and ρ (kg/m^3) is the density of the material. Heat q (W/m^2) and moisture flow g (kg/m^2) through the wall are governed by the differences in temperature, vapour pressure p_v (Pa) and suction pressure s (Pa):

$$q = \underbrace{-\lambda\frac{\partial T}{\partial x}}_{\text{heat conduction}} + \underbrace{g_a \cdot c_a}_{\text{heat convection}} \pm \underbrace{g_v \cdot h_e}_{\substack{\text{heat source due to}\\ \text{evaporation/condensation}}} \tag{25}$$

$$g = g_v + g_a \cdot x_a + g_l = \underbrace{-\delta_p\frac{\partial p_v}{\partial x}}_{\text{vapour diffusion}} + \underbrace{g_a \cdot p_v \cdot 6.21 \cdot 10^{-6}}_{\text{vapour convection}} \pm \underbrace{\lambda_{m,l} \cdot \frac{\partial s}{\partial x}}_{\text{capillary suction}} \tag{26}$$

For air permeable materials the model calculates one-dimensional air flow rate, g_a (kg/m^{2}s), in x-direction and out of predefined pressure differences, and thereby transferred heat and moisture through the materials.

HAM states in the attic, air temperature T_{attic} (°C), moisture content in air v_{attic} (kg/m^3) and air pressure $P_{a,attic}$ (Pa) are found from the energy and mass balance equations:

$$c\rho V_{attic}\frac{dT_{attic}}{dt} = \sum_i Q_i,\ V_{attic}\frac{dv_{attic}}{dt} = \sum_j M_j \text{ and } 0 = \sum_k M_{a,k} \tag{27}$$

where Q_i (W) are heat flows to the attic, e.g. from the roof and the attic floor, ventilation and infiltration, while M_j (kg/s) and $M_{a,k}$ (kg/s) are respective mass flows of vapour and dry air. The first two expressions describe storage of energy and water vapour in air and the last one is for the mass balance of dry air. Heat exchange by radiation is calculated separately and it involves, among others, calculations of view factors between the surfaces. Any spatial variations of air temperature and moisture content in the air are neglected, i.e. a well-mixed air is assumed. Three-dimensionality of the attic is achieved by specifying spatial positions (orientation and tilt) of the one-dimensional components around a ventilated zone, as shown in Fig. 6. However, some local air pressure variations are regarded in the airflow model, as described hereafter.

Transfer and storage properties of building materials, which are all dependent on current temperature, moisture content and air pressure in the materials, can be specified by a user.

As indicated earlier, one of the benefits with the advanced HAM model is a detailed insight in air flows through the attic. While the thermal and moisture models described by Eqs. 24–27 apply only for simulations in the attic, the airflow model (called also A-model, Eq. 27) applies to the whole house, as it is shown in Fig. 7. Therefore, the dimensions of the house are important for these airflow

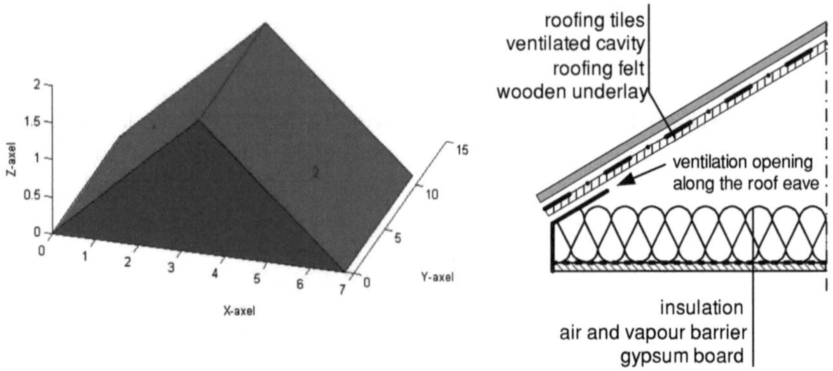

Fig. 6 To the left: The attic zone is defined by the roof elements and the attic floor. To the right: construction of a typical attic in Sweden

Fig. 7 Coupling of a HAM model of an attic with A model of a two-storey house

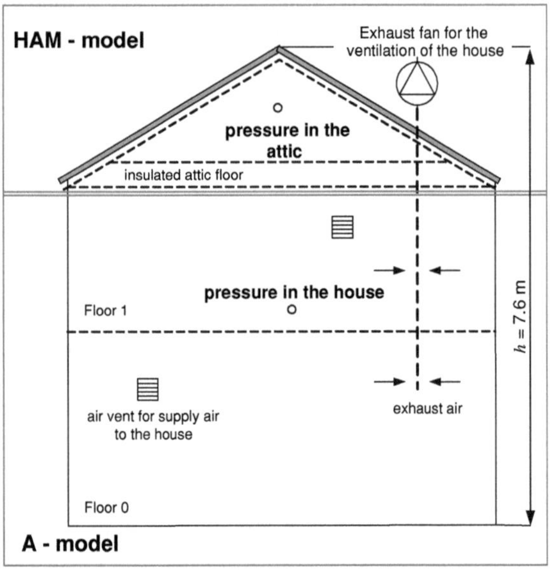

calculations as well as the type, position and airflow characteristics of the components of the ventilation system in the house (a fan and air supply units) and in the attic (ventilation openings along roof eaves). In addition, the distribution and airflow characteristics of the air leakages in the walls, ceiling, gables and roofs must be assumed. Although there is no HAM model of the house, the indoor climate could be determined from the outdoor air temperature according to the methodology in EN 15026 or similar. Alternatively, a full HAM model of the house could be combined with the model of the attic.

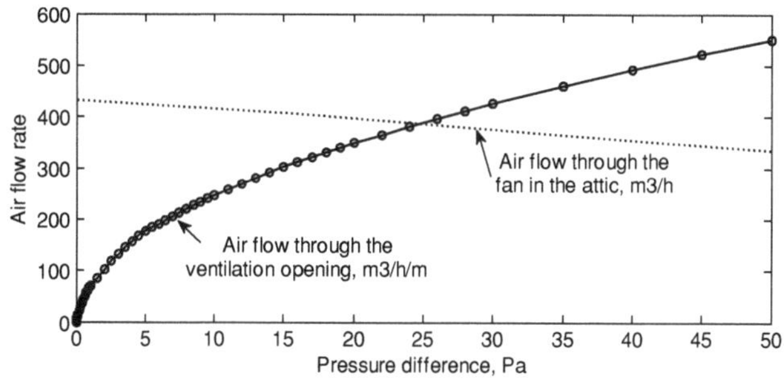

Fig. 8 Airflow rates through the ventilation opening along the roof eave, as measured by Mattsson (2007) and through the fan. The measurements were performed on a gap size of 25 mm between the board protecting the insulation and the roof sheathing and with an insect net. The air flow rate through the fan in the attic is calculated

3.1.2 Ventilation of the Attic

Commonly, the ventilation of the attic is accomplished through openings constructed along roof eaves. The air flow rate through these openings has a nonlinear relationship with the pressure difference across it, as it is found by the measurements of Mattsson (2007) and shown in Fig. 8. The flow characteristic from the figure is implemented as a table with values in the simulations and the values between the points are determined by linear interpolation. For attics where the ventilation is performed through another type of openings, e.g. vents in the gable sides, or mechanically by a fan, the flow characteristics of the openings and fans could be specified by a similar table or by flow equation. Figure 8 shows how the flow characteristic of a fan is implemented in the model as a table with values.

3.1.3 Ventilation and Air Infiltration in the House

Natural ventilation has been the most common method for satisfying the fresh air requirements in domestic buildings in Sweden. It is nowadays largely replaced by mechanical exhaust ventilation. In any case, flow characteristics of ventilation openings and the fan in the house should be specified in order to build an A-model of the house. Besides, it is necessary to specify the flow through air leakages in the building envelope as they are also a part of the ventilation system in the house. For example, in the two-story house from Fig. 7, a ventilation fan together with the vents placed in the walls can create an under pressure of 9 Pa, as it is shown in Fig. 9. When the air leakages are taken into account, the under pressure in the house is about 2 Pa.

The air leakage characteristic of the house is described by the air change rate at 50 Pa, the n_{50} value. This value is adapted and developed in this model using the mean leakage coefficient of the building envelope $\bar{k}$ (m^3·s)/(Pa$^\beta$·m^2):

Fig. 9 Flow characteristics of the fan in the house, air vents and air leakages in the walls and the ceiling

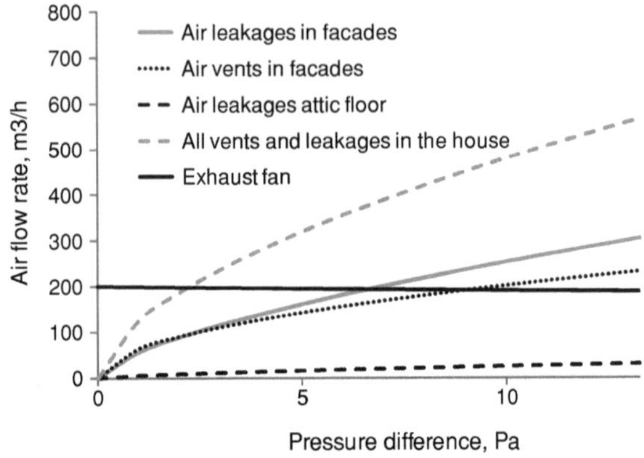

$$\bar{k} = \frac{n_{50} \cdot V}{A \cdot 50^{\beta}} \tag{28}$$

where V is the volume of the house, A is the area that separates the indoor and outdoor climate from each other and β is the exponent of the pressure term, with a typical value of 0.65.

The mean leakage coefficient is usually the only value that can be derived from measurements. Hence, the leakage coefficients of walls, ceiling and the floor should be assumed in simulations. This is done by partitioning the total flow through the house into the flow through the respective part of the building envelope:

$$\begin{aligned} & n_{50} = n_{50,ceiling} + n_{50,floor} + \sum_i n_{50,wall,i} \\ & n_{50,ceiling} = \frac{k_c \cdot A_{ceiling}}{V_{house} \cdot 50^{-\beta}},\ n_{50,floor} = \frac{k_f \cdot A_{floor}}{V_{house} \cdot 50^{-\beta}},\ n_{50,wall,i} = \frac{k_{w,i} \cdot A_{wall,i}}{V_{house} \cdot 50^{-\beta}} \end{aligned} \tag{29}$$

The individual leakage coefficients of the walls, $k_{w,i}$, the ceiling, k_c, and the floor k_f can be found as area-weighted values, following the procedure shown in Essah et. al (2009). With this definition, $k_{w,i}$, k_c and k_f can be changed from one simulation to another but the overall airtightness of the house is the same as long as n_{50} does not change. This approach means that the effects of varying ceiling tightness can be investigated independently of the overall air tightness of the house.

3.1.4 Air Mass Balance in the House and in the Attic

Air pressure inside the house is influenced by wind acting on the house, stack effect, ventilation system in the house and by the air pressure in adjacent zones such as attic. The mean air pressure in the house can be found from the air mass balance Eq. 27. Pressure in the attic is found in a similar way, i.e. by finding the mass balance of all flows through the attic. In this, the stack effect is neglected because of the small difference between the attic and outside temperature.

When only a fan or wind acts on the house, the air pressure differences across the walls and the ceiling are uniform. However, the stack effect leads to two distinctive air pressure zones in the house, above and under the neutral pressure plane (NPP). Consequently, there is non-uniform air pressure distribution over the walls and the ceiling. For the purpose of modelling of the air flow due to stack effect, each wall surface in the model is virtually divided into two sub surfaces, the lower and the upper part (below and above NPP):

$$A_{w,i} = \sum_{j=1}^{2} A_{w,ij} \tag{30}$$

Here, index $j = 1$ is for the wall area below NPP, and 2 when for the area above NPP. The mass air flow rate through area $A_{w,ij}$ in the wall i reads:

$$M_{a,w,ij} = \begin{cases} b_i \cdot A_{w,ij} \cdot k_{w,1} \cdot \left|\Delta P_{a,w,ij}\right|^{\beta_i} \cdot \rho_{a,out} & \text{for air flowing into the house} \\ -b_i \cdot A_{w,ij} \cdot k_{w,1} \cdot \left|\Delta P_{a,w,ij}\right|^{\beta_i} \cdot \rho_{a,in} & \text{for air flowing out of the house} \end{cases} \tag{31}$$

where $\Delta P_{a,ij}$ is the mean pressure difference in the lower and upper parts of the wall, and $\rho_{a,out}$ and $\rho_{a,in}$ are the densities of outdoor and indoor air. The position of NPP changes when wind acts on the house, so thus $A_{w,ij}$, while A_i remains constant. The mass air flow through the ceiling is given in the similar way:

$$M_{a,c,j} = \begin{cases} b_5 \cdot A_{c,j} \cdot k_{w,1} \cdot \left|\Delta P_{a,c,j}\right|^{\beta} \cdot \rho_{a,out} & \text{for air flowing into the house} \\ -b_5 \cdot A_{c,j} \cdot k_{w,1} \cdot \left|\Delta P_{a,c,j}\right|^{\beta} \cdot \rho_{a,in} & \text{for air flowing out of the house} \end{cases} \tag{32}$$

The pressure difference across the walls and the ceiling are found as

$$\Delta P_{i,j} = sign \cdot z_{ij} \cdot \left(\rho_{a,out} - \rho_{a,in}\right) g \tag{33}$$

where *sign* denotes the flow direction, g is the gravity acceleration and z_{ij} is the distance from the position of NPP, as shown in Fig. 10 and specified in Table 4.

The wind induced pressure differences along the walls are found from:

$$P_{wind,i} = C_{wind,i} \frac{\rho_{a,out} \cdot w^2}{2} \tag{34}$$

where $C_{wind,i}$ are the wind pressure coefficients for different facades of a building and w (m/s) is the wind speed at the height of the building ridge. Table 5 shows the wind pressure coefficients for a semi-sheltered low rise building, interpolated from

Fig. 10 When both the wind and the stack effect are present, the NPPs at the windward and the leeward sides of the house are placed on the different heights, $z_{0,windward\ side}$ and $z_{0,leeward\ side}$

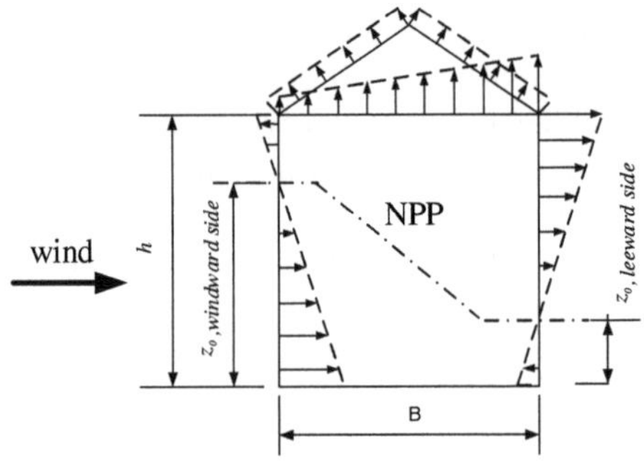

Table 4 Values for *sign* and z_{ij} in Eq. 33

Surface	Height z_{ij}	Sign
Wall, lower part, $j = 1$	$0.5\ z_0$	+
Wall, upper part, $j = 2$	$z_0 + 0.5(h - z_0)$	−
Ceiling, whole area	B	−

B and h are the width/height of the ceiling/wall and z_0 is the distance of NPP from the ground

Table 5 Pressure coefficients on a sheltered low rise building, according Sanders (1996)

Wind angle	0	45	90	135	180	225	270	315
$C_{wind,i}$	0.45	0.2	−0.375	−0.35	−0.225	−0.35	−0.375	0.2

the coefficient given in Sanders (1996) for the buildings in open and urban environments.

Wind speed at the building site (country with scattered brakes) is estimated according to the model given in Sanders 1996):

$$w = 0.52 \cdot h^{0.2} \cdot w_0 \tag{35}$$

where w_0 is the wind speed at 10 m.

The air mass balance equation for the house in reads

$$M_{a,fan} + M_{a,vents} + \sum_{i=1}^{4}\sum_{j=1}^{2} M_{a,ij} = 0 \tag{36}$$

which is an extended form of Eq. 27.

Air pressures in the house and in the attic have to be found iteratively within each time step of simulations. The iteration process is accomplished by two algebraic loops (i.e. one Eq. 27 for the house, the other for the attic), which somewhat slow down the simulation time. For example, one year calculations with the solver ODE23tb and by using variable time steps are performed in app. 4 min on a PC (Intel(R) Core(TM)2 Duo CPU 6600 @ 2.40 GHz). Without algebraic loops, the calculation time can be 10–15 times improved because the model could be run in the acceleration mode.

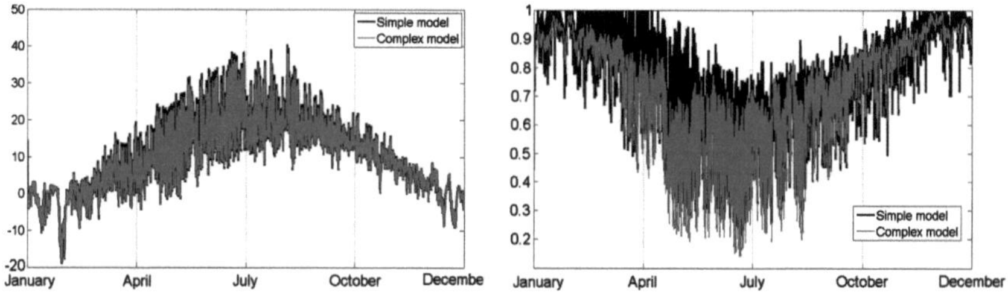

Fig. 11 Temperature and relative humidity at the roof underlay, calculated by the simple and the complex model. Results for year 1979

3.2 Comparison of the Simple and the Advanced Model

By solving ordinary differential Eqs. (3), (9) and (12) in the simple model, and Eqs. 24–27 in the complex model, hourly values of temperatures and relative humidities at internal side of the roof underlay are found. Example of the results obtained is given in Fig. 11. By combining these results with the model for mould growth potential, the probability density function can be calculated.

The air exchange rates are constant for each yearly simulation. The indoor moisture supply for each hour of the year is calculated before the actual dynamic simulation for a year starts. The calculation time is approximately 25 s per sample (one year simulation) when using the simple model and 120 s with the complex model. The given speeds refer to a PC (Intel Core 2 Duo CPU, P9400 @ 2.40 GHz).

By varying weather, ventilation flow rate by outdoor air, air leakage rate from the dwelling below the attic and the moisture excess in the indoor air in the dwelling, six different scenarios are created. Tables 6 and 7 show the probability of exceeding the mould growth potential of one for each of the specified scenarios using 100 yearly samples.

3.3 Alternative Technical Solution for Cold Attics

The results presented in Sect. 2 illustrate the role of ventilation and air infiltration on mould growth risk in attics. The focus in this section is on weather. With the help of the advanced HAM-model, hygrothermal performance of a typical attic construction that is shown in Fig. 6 is studied in four different climates in Sweden: mild and humid west-coastal region, which is represented by the climate of Gothenburg city, mild and humid but somewhat warmer south-west coastal region (city of Lund), mild and dryer east-coastal region (Stockholm) and dryer and colder continental region (Östersund).

The following data are assumed in simulations: the attic is constructed above a two-story single family house such as the one shown in Fig. 7; the roof is pitched

Table 6 The probability of exceeding a mould growth potential of one at the surface of the wooden underlay

	Leaky attic floor $f_{ia} = N(0.2, 0.1)$ (%)	Less tight attic floor $f_{ia} = N(0.1, 0.1)$ (%)	Tight attic floor $f_{ia} = 0$ (%)
Simple model	22	20	12
Complex model	22	18	15

The attic is well insulated and ventilated, $n_e = N(7, 5)$

Table 7 The probability of exceeding a mould growth potential of one at the surface of the wooden underlay

	Leaky attic floor $f_{ia} = N(0.20.1)$ (%)	Less tight attic floor $f_{ia} = N(0.1, 0.1)$ (%)	Tight attic floor $f_{ia} = 0$ (%)
Simple model	25	19	7
Complex model	21	19	8

The attic is well insulated and less ventilated, $n_e = N(2, 1)$

Table 8 The scale for mould index

Mould index	Growth rate
0	No growth
1	Some growth detected only with microscopy
2	Moderate growth detected with microscopy (coverage more than 10 %)
3	Some growth detected visually
4	Visually detected coverage more than 10 %
5	Visually detected coverage more than 50 %
6	Visually detected coverage 100 %

at 30° and the roof sides face north and south; the area of the ceiling is 75 m^2; 400 mm thick mineral wool insulation is laid on a horizontal plasterboard ceiling (attic floor) and with an air barrier (plastic foil) in between; 29 mm thick underlay, separating loft space from the external weatherproof roof covering (usually tiles), is made of tongued and grooved timber laths; the top side of the underlay is protected from rain and moisture by roofing felt; the attic is ventilated through 20 mm wide openings constructed along the roof eaves and the ventilation is done by outdoor air; n_{50} for the house below the attic is 2 1/h and 9 % of that (0.18 1/h) comes from the leakages in the ceiling. In this way, the attic and the house represents a modern single-family house in Sweden.

The risk for mould growth is estimated from mould index, a method developed by Hukka and Viitanen (1999). Table 8 shows the grading system of mould index. Essentially, only the values less than 1 indicate very low or no risk for mould

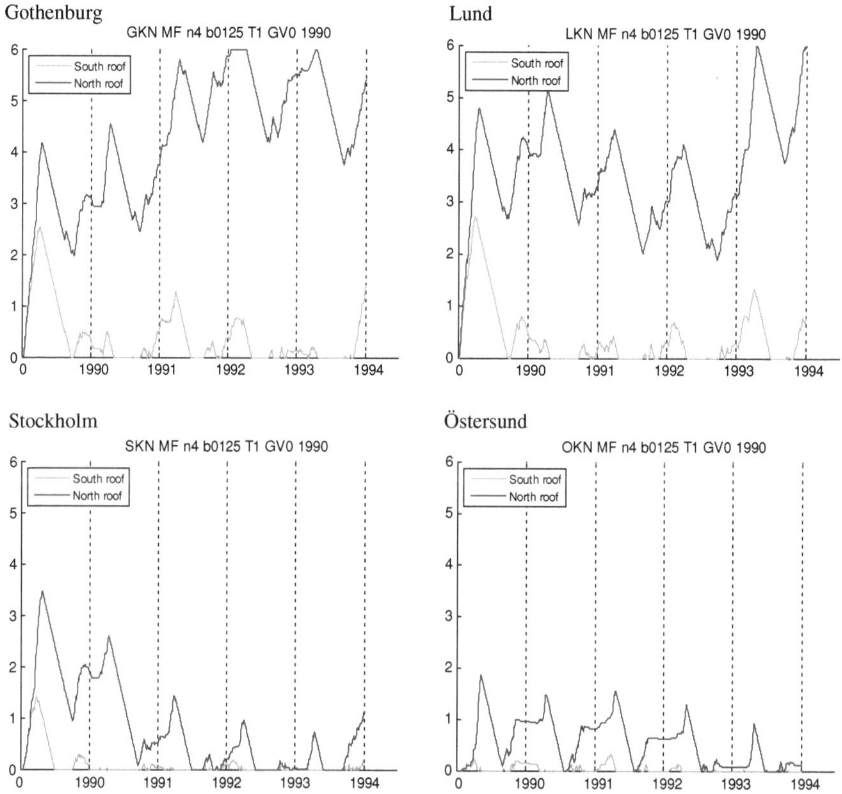

Fig. 12 Mould index for the roofs underlay in a typical attic, for four different locations/climates. Results from the advanced HAM-model of an attic

growth. In practice, the mould is visible firstly when mould index reaches value 3 or higher.

Based on the calculated temperatures and relative humidities on internal sides of the roof, the mould growth index is found for each side of the roof and for five consecutive years. The results of calculations, which are shown in Fig. 12, indicate that the risk for mould growth is highly dependent on the climate. The highest risk is found in mild and humid climate of Gothenburg and the lowest in dry and cold Östersund. For the latter, this particular attic construction could be regarded as risk-free even for the north roof side, which is in all other climates remarkably more vulnerable than the south side.

In search for an absolute mould risk-free attic construction, the building practitioners in Sweden have shown interest in alternative solutions for cold attics. Calculated hygrothermal performance for one of the alternatives, which involves mechanical ventilation and a relatively air tight attic construction, as depicted in Fig. 13, has shown promising results. Based on calculations, the mould index for

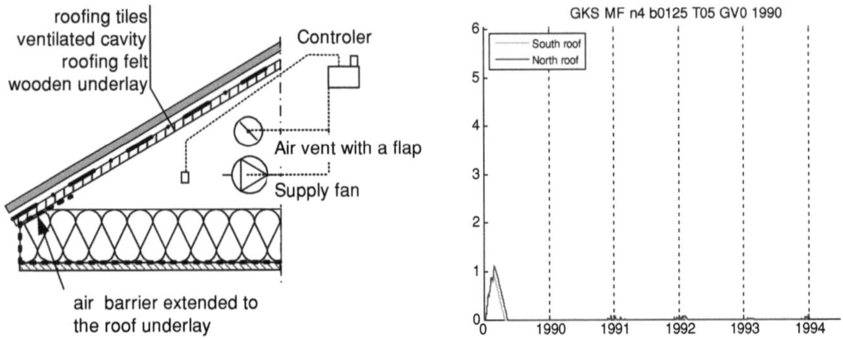

Fig. 13 To the left: Mechanically ventilated attic with a fan, an air vent and a controler. To the right: Daily mould growth index for hygrothermal conditions inside the mechanically ventilated attic placed in Gothenburg

the roofs underlay of an attic placed in Gothenburg is practically 0. The small peak that is present in the beginning of simulations indicates the influence of in-built moisture. Note that all other conditions are exactly the same as for the typical attic. This alternative construction is a result of a patented solution for attic ventilation (Hagentoft et al. 2008). More details about the ventilation system and the validation tests for this attic construction are given in the next section.

4 Adaptive or Controlled Ventilation

To get optimized ventilation, whatever type of external climate, attic insulation, airtightness etc. ventilation must be controlled and adopted to the present situation. By using sensor technology, mechanical ventilation and making the attic as air tight as possible this can be achieved; see Fig. 14. A basic system would comprise mechanical fans and dampers controlled by attic and outdoor climate sensors installed in a sealed attic without vents. The ventilation system runs only when the outdoor air has a potential to dry out the attic.

4.1 Field Studies of One-Family Houses in Stockholm

Installation of measuring equipment and controlled ventilation were made during February 2008. Measurements from autumn 2008 onwards are given below in Figs. 17, 18, 19 and 20. Figure 15 shows the position and orientation of the houses. Figure 16 shows photographs of the houses with adaptive ventilation i.e. Test houses 12 and 22.

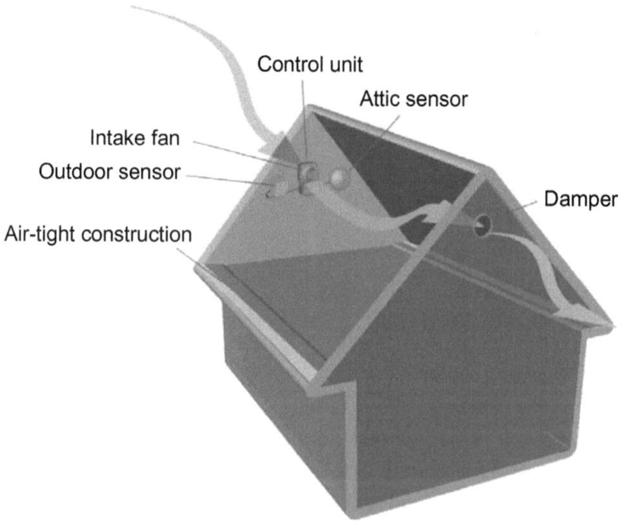

Fig. 14 Adaptive ventilation of a cold attic

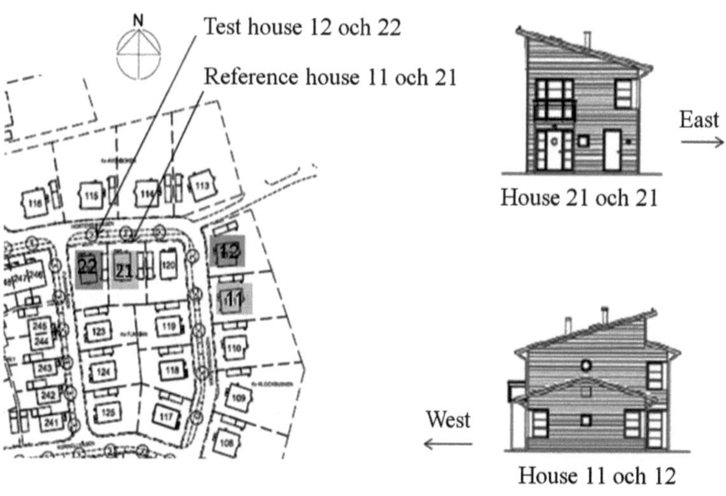

Fig. 15 Position and orientation of the houses in Stockholm. The *red-colored houses*, 12 and 22 have adaptive ventilation while, the *blue*, 11 and 21 are naturally ventilated

All four houses have two floors. Houses 11 and 12 have a floor area of 130 m^2 while House 21 and 22 have the floor area 121 m^2. The houses have pitched roofs, which is built from the inside starting with 21 mm wooden T&G board, base board YAP 2200, litter and batten and counter batten (25 × 38 mm) and roof tiles. All houses have mechanical exhaust ventilation. The cold attic floor is insulated with 450 mm loose fill insulation. The reference house is ventilated by open slots along

Fig. 16 Photographs of test houses, 12 and 22, with controlled ventilation

the eaves. The research projected was coupled to the building project a bit too late in the construction project planning to get the airtightness of the attics properly installed for the test houses and thus the best possible solutions could not be realized. Air sealing was done using corrugated paper board (in some places light could be seen leaking in from outside). In the attic of Test house 12, non-woven fibre sheeting applied on the exterior wood panelling constituted the air tightening.

The test houses have served as demonstration houses during the measurement period and were not permanently occupied. To ensure that the moisture load inside these houses were correct (i.e. not too low), humidifiers were used giving 40 % RH. Reference houses have been occupied during the entire measurement period.

4.1.1 Initial Status of Buildings

Air tightness measurements at 50 Pa were carried out by the building contractor (NCC) and are available for the two test houses (the occupied interior space). For houses 12 and 22 the air leakage was 0.3 l/m^2 s and 0.7 l/m^2 s respectively. Available data for six houses in the area give a spread between 0.3 and 0.8 l/m^2 s.

Measurement of temperature and relative humidity are made at two places on the test attics as well as externally at the north gable. One sensor is placed centrally in the attic and the other further down towards the eaves at the lower part of the roof. Sensors were placed on the rafters in close proximity to the wooden board surface. For the reference houses only centrally placed sensors were used.

Air tightness measurements of the experimental attics were carried out using the existing fan for the adaptive ventilation in the attic. Since the fan capacity is not sufficient to reach 50 Pa, extrapolations were made, based on the premise that air flow is proportional to the square root of pressure difference and the effective aperture area. The excess pressure was not stable so a range is specified in Table 9. The table also indicates the measured moisture content at the time of installation.

The targeted air exchange rate at 50 Pa was less than 10 1/h (preferable less than 7 1/h).

Table 9 Moisture ratio in the cold attics at time of installation, 13th of February, and estimated air exchange rates at 50 Pa pressure

House	Ventilation	Moisture ratio (%) Wooden board	Moisture ratio (%) Rafters	Air exchange rate, 1/h at 50 Pa
11	Natural	19	16.5	–
12	Controlled	19.6	15.6	20–30
21	Natural	–	–	–
22	Controlled	19.9	17.8	40–55

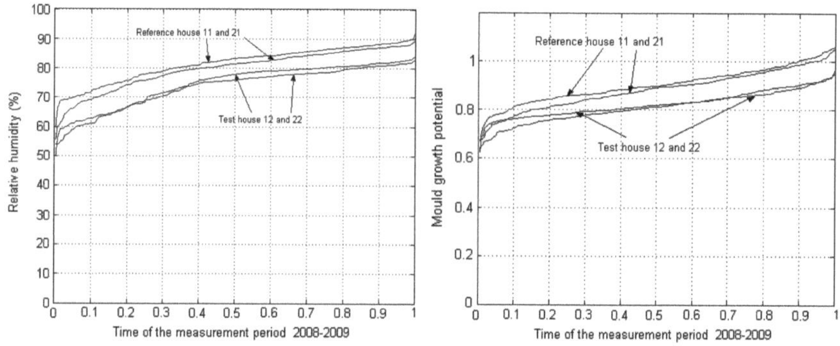

Fig. 17 To the left: Duration graphs for the measured relative humidity (to the left) and the mould growth potential (to the right) in the cold attics with adaptive ventilation. Test houses 12 and 22, and attics with natural ventilation, Reference houses 11 and 21

4.1.2 Measurement Results

Measurement results are shown in Fig. 17 below in terms of duration graphs for the relative humidity for the period from August 2008 until March 2009. Mould growth potential on average during the measurement period for both test houses are 0.82 while for the reference houses it is 0.9 (Reference house 11) and 0.88 (Reference house 21).

An interesting observation made from measurements is that the humidity by volume in the cold attics is on average lower than outside during the measurement period. There is thus a negative moisture supply in the cold attics (as compared to the outside conditions). The reason for this can be both explained by moisture absorption in wood material in the attics and the adaptive ventilation. For cases with adaptive ventilation, the negative moisture supply is 0.5–0.8 g/m^3, while the attics with natural ventilation 0.2–0.3 g/m^3.

Moisture in rafters and in the wooden board was measured in the test houses the first of April 2009. For the rafters, the surface moisture ratio (by weight) varied in the ranges; Test house 12: 11,9–12,1 %, Test house 22: 11,3–12,0 %. In the board, the measured interval was; house 12: 11,8–13,5 %, houses 22: 11,1–12,9 %.

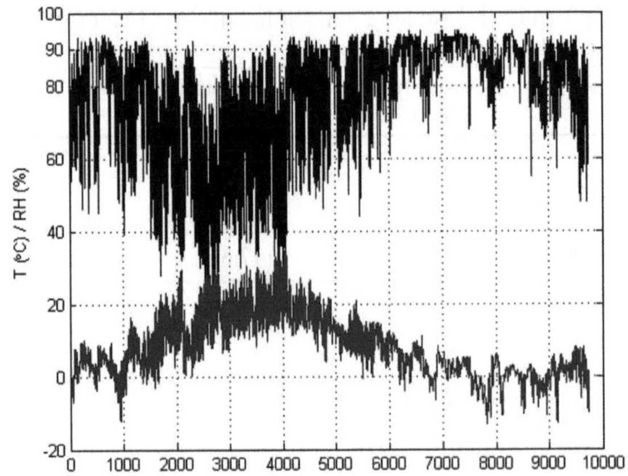

Fig. 18 Measured outdoor air temperature and relative humidity. Time in hours between 14/2/2008 and 25/3/2009

The measurements show a trend toward slightly higher moisture content in the board compared with rafters.

Figure 18 shows the measured outdoor air temperature and relative humidity during the measurement period. The measured RH and temperature inside the attics are shown in Figs. 19 and 20.

4.1.3 Field Studies of Multi-Family Houses in Gothenburg

The adaptive ventilation was in operation in early autumn 2008. Four separate spaces were ventilated. The cold attics were planned using the adaptive ventilation design, which means that a much better air sealing can be expected than in the previous presented test houses. Unfortunately no attic air tightness has been measured. The cold attic floor is a timber framed one with a vapour barrier and 500 mm loose-fill insulation (Fig. 21).

Figure 22 shows the measured relative humidity and the mould growth potential in the four ventilated cold attics.

4.2 Comparisons Between Measurements and Simulations: One-Family House in Stockholm

Data for temperature and RH of the outdoor air is taken directly from the measurement data on site. Solar data from SMHI (met. station) and long wave radiation from the computer program Meteonorm. These last data are considered to be the most uncertain.

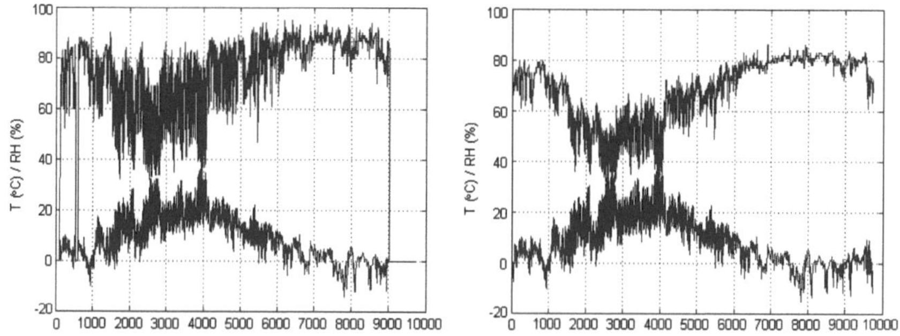

Fig. 19 Measured air temperature and relative humidity in Reference house 11 (*left*) and Test house 12 (*right*). Time in hours between 14/2/2008 and 25/3/2009

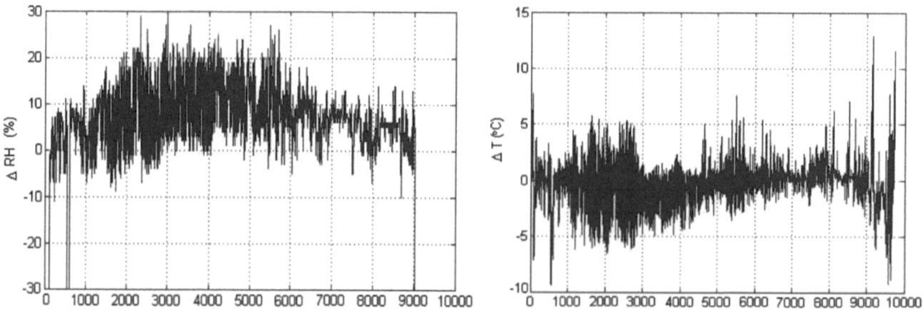

Fig. 20 Differences in relative humidity (*left*, i.e. RF_{11}–RF_{12}) and temperature (*right*, T_{11}–T_{12}) between Reference house 11 and Test house 12. The average differences over the period are +7.4 %. and −0.13 °C. Time in hours between 14/2/2008 and 25/3/2009

Fig. 21 The Harbor house in Gothenburg, Sweden (photo: Staffan Bolminger Älvstranden Development AB)

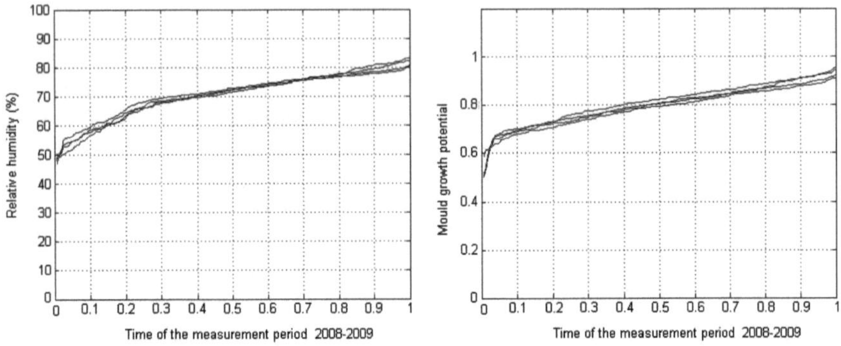

Fig. 22 Duration graphs for the measured relative humidity (to the left) and the mold growth potential (to the right) in the four cold attics of the Harbor house

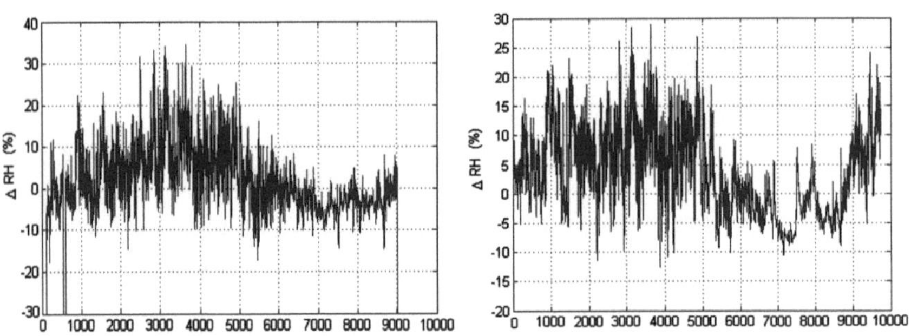

Fig. 23 To the left: Difference in measured and calculated relative humidity in reference house 11 (i.e., $RF_{11,measured} - RF_{11,calc}$). The average difference over the period is: 2.1 %. To the right: Difference in measured and calculated relative humidity in test house 12 (i.e., $RF_{12,measured} - RF_{12,calc}$). The average difference over the period is: 3.7 %. Time in hours between 14/2/2008 and 25/3/2009

The ventilation slots for the naturally ventilated attics are 20 mm wide and has a length of 7.2 m. For the attics with controlled ventilation, the width of the slots is assumed to be 2 mm, based on the air tightness and assuming that Dick's formula is valid. The house internal air volume is 349 m^3 and the area of the building envelope is 239 m^2. The calculations assumed that 24 m^3/h of air leaking out of from the attic at 50 Pa. The attic floors are considered to be very air tight and small variations in the tightness changes calculation results only marginally. Exhaust ventilation in itself reduces leakage upwards which makes the situation relatively insensitive to air tightness. A under pressure in the living space, compared to the exterior, tends to move the air from the attic to the living space instead of the opposite direction. The calculations show a slightly lower average RH (2–4 %), Fig. 23, and warmer (0.8–0.9 °C), Fig. 24, than the measurements of the attics. On

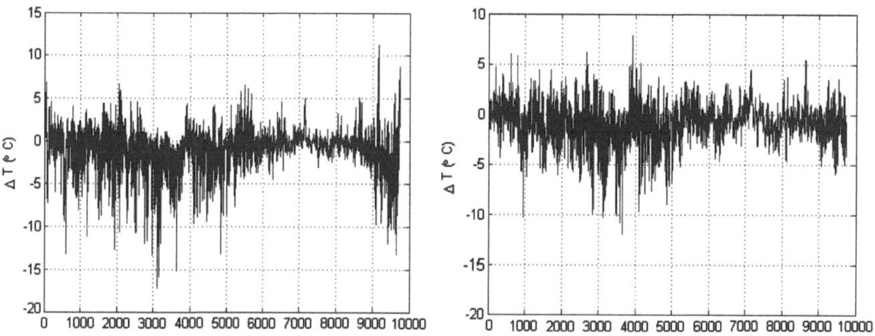

Fig. 24 To the left: Difference in measured and calculated temperature in reference house 11 (i.e., $T_{11,measured} - T_{11,calc}$). The average difference over the period is: −0.9 °C. To the right: Difference in measured and calculated temperature in test house 12 (i.e., $T_{12,measured} - T_{12,calc}$). The average difference over the period is: −0.8 °C. Time in hours between 14/2/2008 and 25/3/2009

Fig. 25 Difference in calculated relative humidity between reference house 11 and test house 12 (i.e., $RF_{11,calc} - RF_{12,calc}$). The average difference over the period is: 8.6 %. Time in hours between 14/2/2008 and 25/3/2009

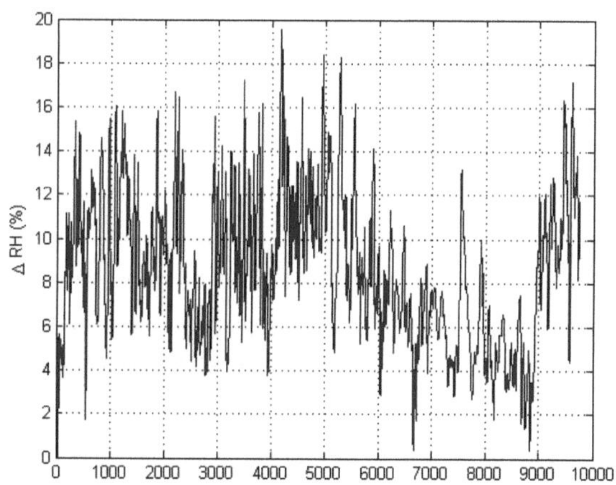

an average the measured RH tends to be higher than the calculated ones during the summer period and lower during the winter period.

At the computational comparison between natural and controlled ventilation for reference house 11 and test house 12, Fig. 25, an RF difference of 8.6 % is obtained.

The calculations show, contrast to measurements, on medium-RF (−1.9–0.3 %); i.e. slightly higher RF values. Calculations give a difference (0.1–0.8 °C), i.e. lower temperatures than measured in the attics. At the computational comparison between natural and controlled ventilation for reference house 12 and test house 22 an RF difference of 8.7 % is obtained.

5 Discussion and Conclusions

Cold attics are known to be risky constructions in Sweden concerning moisture safety and mould growth. Hygrothermal analyses of cold attics are demonstrated both using simplified and advanced models. The advanced model includes detailed air pressure distributions and air leakages as well as accounting for the orientation of the building. The latter is important in order to include the influence of the position of vents and solar radiations on different parts of the roof. For instance the north facing roof is shown to be more susceptible for mould growth. Nevertheless, for low sloped roofs the simplified and advanced models show good agreement for the probability distribution of mould growth. Furthermore, the probability distribution confirms and is in consistency in terms of magnitude with the observed risk of mould growth.

Field measurements of four cold attics in the Stockholm area validate the advance simulation model. It also shown that adaptive ventilation provides more stable and lower relative humidity levels during the winter than the cold attics with traditional ventilation design.

The advanced model show that the local climate is of great importance for the hygrothermal condition and mould risk of attics. The highest risk is found in mild and humid climate and the lowest in dry and cold climate.

References

Ahrnens, C., Borglund E.: Fukt på kallvindar–En kartläggning av småhus i Västra Götalands län, Examensarbete 2007:11, Institutionen för bygg- och miljöteknik, Chalmers tekniska högskola. (In Swedish) (2007)

Arfvidsson, J., Harderup, L.-E.: Moisture safety in attics ventilated by outdoor air. 7th symposium in building physics, Reykjavik, Island (2005)

Essah, E., Sanders C., Baker, P., Sasic Kalagasidis, A.: Condensation and moisture transport in cold roofs: effects of roof underlay. Build. Res. Inform. **37**(2), 117–128 (2009)

Hagentoft, C. E.: Textbook: introduction to building physics. Studentlitteratur, ISBN 91-44-01896-7 (2001)

Hagentoft, C.-E., Sasic Kalagasidis, A., Adl-Zarrabi, B., Roels, S., Carmeliet, J., Hens, H., Grunewald, J., Funk, M., Becker, R., Shamir, D., Adan, O., Brocken, H., Kumaran, K., Djebar, R.: Assessment method of numerical prediction models for combined heat, air and moisture transfer in building components. Benchmarks for one-dimensional cases. J. Therm. Envel. Build. Sci. **27**(4), 327–352 (2004)

Hagentoft, C.E., Sasic Kalagasidis, A., Nilsson, S.F., Thorin, M.: Mould growth control in cold attics through adaptive ventilation. In: 8th Symposium on Building Physics in the Nordic Countries, Copenhagen, June 2008

Hagentoft, C.-E., Sasic Kalagasidis, A.: Mould growth control in cold attics through adaptive ventilation. Validation by field measurements. In: 12th International Conference on Performance of the Exterior Envelopes of Whole Buildings, 2010, Clearwater Beach, Florida

Holm, A., Lengsfeld, K.: Hygrothermal performance of unfinished attics (ventilated roofs)—an experimental study. Research in building physics and building engineering. In: Proceedings from the Third International Building Physics Conference, Montreal

Hukka, E., Viitanen, H.A.: A mathematical model of mould growth on wooden material. Wood Science and Technology 33, Springer (1999)

IEA, Annex 14. Condensation and Energy, Catalogue of Material Properties Vol 3, IEA-ECBCS. K.U Leuven, Belgium 1992

Mattsson, B.: Studies on building air leakage—a transient pressurisation method, measurements and modelling. Ph.D. thesis, Department of Civil and Environmental Engineering, Division of Building Technology, Chalmers University of Technology, Gothenburg (2007)

Samuelson, I.: Hygrothermal performance of attics. In Swedish. SP Report 1995:68, Sweden. (In Swedish) (1995)

Sanders, C.: IAE Annex 24, Final Report, vol. 2. Environmental conditions. KU Leuven, Belgium, Laboratorium Bouwfysica (1996)

Sanders, C.H.: Modelling condensation and airflow in pitched roofs—building research and establishment (BRE) information paper, IP 05/06. BRE Press, Garston, Watford—UK. ISBN 1-86081-912-5, pp. 1–7 (2006)

Sasic Kalagasidis, A.: HAM-tools. An integrated simulation tool for heat, air and moisture transfer analyses in building physics. Doctoral thesis, Chalmers University of Technology, Sweden (2004)

Woloszyn, M., Rode, C.: ECBCS Annex 41. final report 1: modelling principles and common exercises. Available on www.ecbcs.org (2008)

Åberg, O.: Moisture and mould problems a threat against sustainable buildings. The BETSI study in Sweden. Available on www.boverket.se (http://www.boverket.se/Sok/?quicksearchquery=moisture+and+mould, 2011-10-17) (2011)

Controlled Ventilation of Historic Buildings: Assessment of Impact on the Indoor Environment via Hygrothermal Building Simulation

Florian Antretter, Sarah Kosmann, Ralf Kilian, Andreas Holm, Fabian Ritter and Barbara Wehle

Abstract Historic buildings, when they are unheated, often face problems of summer condensation. After the cold season warm, humid air enters the building and condenses on the walls, a problem that can occur during the entire warm season. This leads to moisture related problems such as mould or algae growth on building surfaces. High humidity can also damage works of art inside the building. One possibility to lower the level of relative humidity is to ventilate the building every time when the dewpoint outdoors is lower than indoors. On the other hand, the resulting humidity fluctuations may also cause damages on works of art and bringing in cold air will further lower the temperatures of the whole building and its walls thus being counter productive to the drying process. An automatic system for ventilating historic, unheated buildings is assessed using building simulation software (WUFI®Plus) on two case studies, the St. Margaretha church in Roggersdorf, near Holzkirchen, Germany and the Gatehall of Lorsch, Germany.

F. Antretter (✉) · S. Kosmann · R. Kilian · F. Ritter · B. Wehle
Fraunhofer-Institut für Bauphysik, Holzkirchen, Germany
e-mail: florian.antretter@ibp.fraunhofer.de

S. Kosmann
e-mail: sarah.kosmann@ibp.fraunhofer.de

R. Kilian
e-mail: ralf.kilian@ibp.fraunhofer.de

F. Ritter
e-mail: fabian.ritter@ibp.fraunhofer.de

B. Wehle
e-mail: barbara.wehle@ibp.fraunhofer.de

A. Holm
University of Applied Sciences, Munich, Germany
e-mail: Holm@fiw-muenchen.de

V. P. de Freitas and J. M. P. Q. Delgado (eds.), *Hygrothermal Behavior, Building Pathology and Durability*, Building Pathology and Rehabilitation 1,
DOI: 10.1007/978-3-642-31158-1_5,

Keywords Historic building · Controlled ventilation · Hygrothermal whole building simulation · Summer condensation

1 Introduction

A high number of unheated historic buildings all over Europe show problems with mould or algae growth. This is very often due to summer condensation. Several options to reduce the risk for mould or algae growth exist. One is, to dehumidify the air to keep the relative humidity inside the buildings on a low level. But dehumidification leads on one hand to a high energy demand; on the other hand it is very often not possible in the affected buildings due to missing space for the equipment and/or missing power supply for operation. Another option is using a controlled ventilation system. Controlled ventilation systems measure the absolute humidity inside and outside the building. If the absolute humidity outside is lower than inside, there is potential to dry the building by an increased ventilation.

In this article two buildings in Germany are assessed. Both show moisture related problems. A hygrothermal whole building simulation software, WUFI®-Plus, is used to assess the building temperature and humidity conditions with different ventilation strategies.

2 Humidity Control in Historic Buildings and Hygrothermal Whole Building Simulation

Adaptive ventilation is used to lower the relative humidity in buildings. The aim is usually to reduce the risk for biodeterioration. Adaptive ventilation means to ventilate only when the exterior absolute humidity level is lower than the interior absolute humidity level. This way humidity can be removed from the inside. On the other hand it is important to stop the ventilation if the exterior absolute humidity is higher than the interior to avoid an additional moisture supply into the building. Furthermore it is important to also measure the temperatures and stop ventilation if the exterior temperature is too low to reduce an additional cooling effect or freezing due to ventilation with exterior air. Thus a mechanical fan and damper with temperature and humidity sensors attached to a control unit are required to conduct controlled ventilation.

For the assessment of mould growth the times where a combination of temperature and relative humidity prevail are critical (Sedlbauer 2001). As a general rule of thumb the relative humidity should not exceed 80 % RH at room temperature. Algae growth usually requires liquid water on the surface. This means, that the dew point temperature on a surface is higher than the actual temperature.

But not only mould and algae growth are critical for the assessment of historic buildings, also the fluctuations of the conditions, especially of relative humidity, is important. According to [Holmberg 2001] the daily relative humidity fluctuations should be kept below 15 % RH.

Broström et al. (2011) assessed the application of adaptive ventilation on a case study building in Gotland, Sweden. They concluded that the adaptive ventilation has had a significant drying effect and that the mould risk was kept at an acceptable level for most of the times. Even the short term RH variations were acceptable. But they also mention that further research is required to understand the effects of hygrothermal inertia and the integration of heating and dehumidification.

These parameter variations can be studied by using hygrothermal whole building simulation software. The advanced hygrothermal whole building simulation tool WUFI®Plus is used for simulation (Holm et al. 2003). This software couples whole building energy modeling with hygrothermal component modeling. It calculates the coupled heat and mass transfer, i.e. mass transfer by diffusion and liquid water transport, for every enclosure assembly and couples heat and moisture fluxes from the inner surfaces with the zone temperature and relative humidity. This model allows the combined assessment of hygrothermal conditions of the building envelope, indoor climate and energy demand of the building. It requires in general hourly exterior climate data, which is provided by the climate database that comes with the software or by measured data. Materials are selected from a material database or specified by providing thermal and hygric properties, so that assemblies can be built from one or more layers of different materials. As the indoor climate is a result of the simulation, appropriate assumptions for inner sources and set-points need to be made. The set-points can be met by specifying an ideal HVAC system. A predefined ventilation control algorithm is available that allows the application of an adaptive ventilation strategy.

3 Case Study Buildings

For the simulation of the hygrothermal building behaviour the software tool WUFI®Plus was used (Holm et al. 2003). In the following the general conditions, the assumed building geometry and materials as well as the heating, ventilation and air conditioning operations are described.

3.1 Church of Roggersdorf

After a complete renovation of the chapel of ease St. Margaretha (see Fig. 1) in September 2004, the churchwarden again noticed moisture damage on the walls. A subsequent climate measurement showed that the damage was a matter of condensation that occurs mainly in the transitional period during spring-time. At that

Fig. 1 Picture and screenshot of the computer model of the exemplary church

time of the year the building is still cold because of the winter. If warm, humid air enters the building due to natural air flow or uncontrolled ventilation, it condenses on the cold wall surfaces. But even in the summer and fall warm humid weather conditions can cause condensation problems.

From December 2004 to August 2006 the temperature and relative humidity were measured inside and outside the church. To asses condensation events the wall surface temperature was also measured on the Western wall at the joint to the floor. Weather data was available from the Fraunhofer IBP outdoor testing facility at Holzkirchen, only 5 km away from Roggersdorf. During the period when measurements were taken the climate inside the church showed a high average humidity with values over 75 % RH for more than half of the time of the year (see Fig. 2). During winter time freezing conditions with temperatures below 0 °C are recorded for more than six weeks in a row (Kilian 2007). Subsequently, the church starts with very low wall temperatures into spring time and the warm season.

Another problem that became evident was the uncontrolled opening of windows and doors over the warming period of spring that was supposed to reduce the moisture levels in the space by ventilating. In the summer 2005 "ventilation traffic lights" showing the times when the water content of the outdoor air was lower than indoors were used to give advice to the warden of the church. This significantly increased the daily fluctuations of RH; in the spring and summer 2005 the fluctuations were above 15 % RH for more than 30 days. Daily changes above 15 % RH are thought to be critical to works of art (Holmberg 2001), as the risk of structural damages due to swelling and shrinking of materials increases with the range of RH change per day. Also the newly restored altarpiece from the nineteenth century started to show additional damage to the gilding.

As a consequence, the installation of a controlled ventilation system was discussed with the church authorities. Such systems permit to compare the absolute humidity inside and outside the building. Whenever the outer humidity is lower, a ventilator starts bringing dryer air into the building. Also, boundary conditions for the lower temperature can be given, as well as a maximum allowed range of RH change in the last 24 h. Practical experience has shown that this type of ventilation

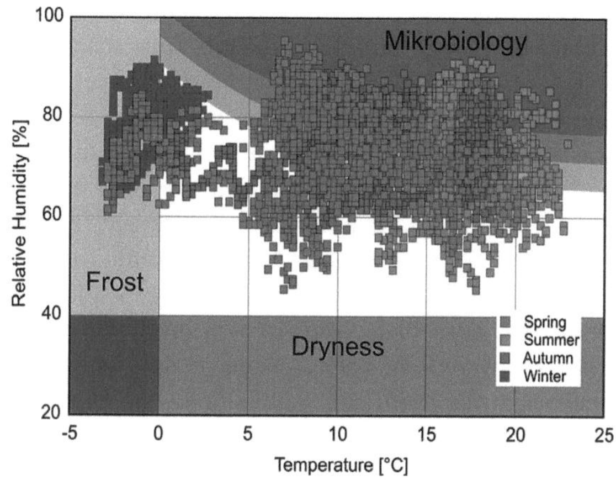

Fig. 2 Temperature and humidity in the church during 2005. Half of the time RH lies above 75 %, during winter the indoor temperature falls below freezing conditions for 6 weeks

system can help reduce the RH range and reduce the overall moisture in a room over the year, as was done for the church located in Urschalling, Bavaria (Künzel 2010). However, the open questions are: how good is the system for the respective location; what can it achieve; how much energy does it use; and, what is the overall impact when bringing colder air into the building during winter and in summer nights? To assess the possibilities and limitations of these systems, whole building simulation was used to predict the indoor environment in the church using weather data from Holzkirchen.

3.1.1 Boundary Conditions

The time of calculation lasted from January 1st to December 31st of the year 2005. The simulation was carried out in time steps of 5 min to reproduce a realistic ventilation control. Given that the location of the church was in Roggersdorf, which is located near Holzkirchen, the meteorological data of Holzkirchen for the year 2005 was used for the simulation. A weather file was created which consisted of 5-min values of exterior temperature, relative humidity (both in Fig. 3), global and diffuse radiation, rain, wind and barometric pressure. Additionally a file with ground temperatures at a depth of 0.5 m in the soil was generated to take into account conditions below the foundation of the church. Inner loads were not taken into account for the base case, assuming that there is no church service.

3.1.2 Building Geometry and Material Data

A detailed model of the chapel's geometry was created in WUFI®Plus (Fig. 1). The main building is about 12 m long and 9 m wide. The height from the ground to the top of the roof is about 6 m and to the top of the tower about 9 m. For the simulation the main body was divided into three zones where the climate was modeled: the nave, sacristy and the attic. A mechanical ventilation system in the nave will provide outside air to this zone with an air exchange rate of 5 h^{-1}, as long as the outside absolute humidity is lower than the absolute humidity inside and the exterior temperature remains above 0 °C. The infiltration ventilation air was taken from outside, with an infiltration air exchange rate of 0.4 h^{-1} for the nave. The tower as well as the entrance area was treated as an attached unheated zone with assumed exterior temperature and relative humidity in the zones.

The outer walls of the chapel are built from sandstone with a thickness of 0.52–0.77 m. At the inner surfaces there is a layer of lime plaster which is 2 cm thick. The base plate consists of a layer with loose material (0.15 m) which is mainly covered by natural stone plates with a thickness of 5 cm. Under the benches at the north and south side of the church there are panels of hard wood instead. The ceiling to the attic has an overall thickness of 0.3 m consisting of 2.5 cm softwood, a 15 cm thick air layer, 10 cm mineral wool and again 2.5 cm softwood. The chapel's roof is mainly made of softwood with shingles on the outside. The windows have a single glazing with an overall thermal transmission value of 3.7 W/m^2K.

3.1.3 Parameter Variations

The simulations were run with different parameter variations. The base case is a simulation as described in the previous section. The simulation results of this variant are always compared to a variant without any ventilation control, i.e. only natural ventilation mainly by infiltration. Further simulations were run with air change rates of 1 h^{-1} and 10 h^{-1} to show the influence of the capability of the ventilation system. A further simulation was run where only the energy transport was modeled whereas the coupled equations for heat and moisture transport in the building envelope were not modeled. This simulation allowed assessing the necessity of a simulation model capable of modeling the coupled transport equations.

3.1.4 Results

Figure 3 shows the modeled temperature and relative humidity inside the nave without additional ventilation and with an assumed air change rate of 5 ACH for ventilation during all periods with lower exterior absolute humidity than inside and while exterior temperatures are above 0 °C. The inner climate conditions are compared to the outer climate conditions.

Expectedly lower fluctuations for temperature as well as for relative humidity are found for the interior conditions compared to the outside climate. Not only the peaks are damped, but also in longer periods with very low exterior temperatures the inner temperatures do not reach the mean low temperature. The fluctuations are lowest in the model without additional mechanical ventilation. Especially in spring,

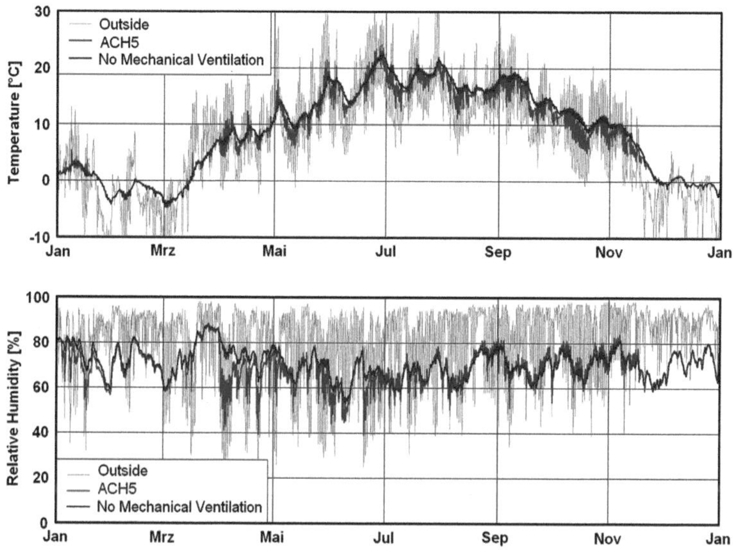

Fig. 3 Temperature and relative humidity outside and modelled conditions inside the church during the whole simulation period with (Air Exchange 51/h) and without mechanical ventilation

the relative humidity is reduced in the controlled ventilation case, but also the temperature inside the church. No effect is found in the cold month due to the temperature limitation at 0 °C of the ventilation control.

The modeled ventilation control acts on absolute humidity. A comparison in Fig. 4 for absolute humidity inside and outside shows also a dampening of the fluctuations from outside to inside for the case without ventilation. The general trend is followed pretty close. The above mentioned effectiveness of the ventilation control in the cold month is reflected by the related ventilation status. Almost no ventilation actions are performed from mid November until end of March. In the warm months the peaks in exterior absolute humidity appear reduced at the inside. The benefits of ventilation control become evident by comparing the absolute humidity with and without controlled ventilation from April to November.

A closer look to the absolute humidity conditions in October is given in Fig. 5; details are provided of the resulting higher fluctuations of absolute humidity inside the building because of active ventilation control. But it also clearly shows that lower interior absolute humidity's are achievable with effective ventilation during periods with lower exterior humidity.

A variation of different air change rates as well as a simulation without moisture buffering—standard building energy simulation—was performed and the results for October are shown in Fig. 6. An improvement in efficiency is achievable by increasing the air change rate during control action from 1 to 5 ACH. In the modeled case, a further increase up to 10 ACH does not further reduce the absolute humidity indoors, as the exterior minimum absolute humidity is also reached with

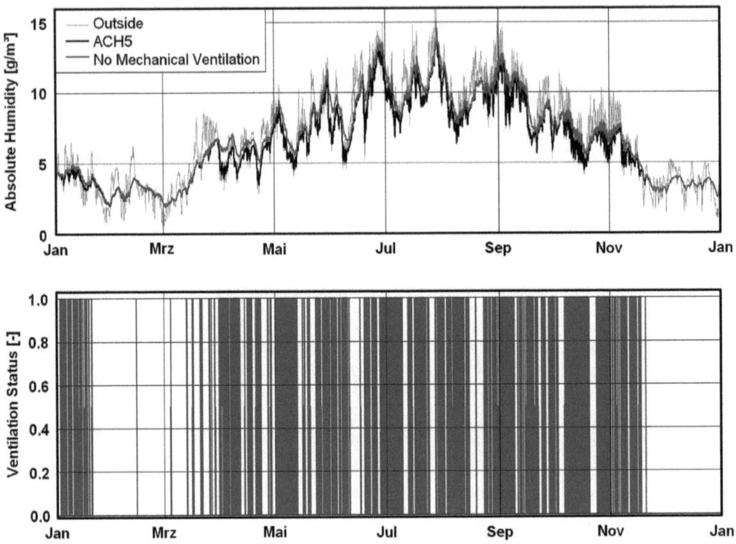

Fig. 4 Absolute humidity outside and inside the church as well as ventilation status during the whole simulation period

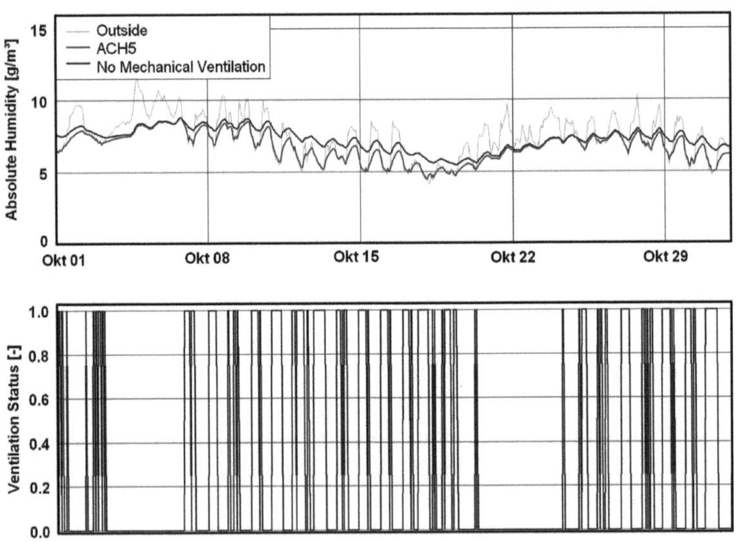

Fig. 5 Absolute humidity outside and inside the church and ventilation status for simulation in October

5 ACH. In the case of a simulation without modeling the coupled heat and mass transfer in the building envelope but using only a mass balance for interior humidity, the interior absolute humidity follows very closely the exterior absolute humidity.

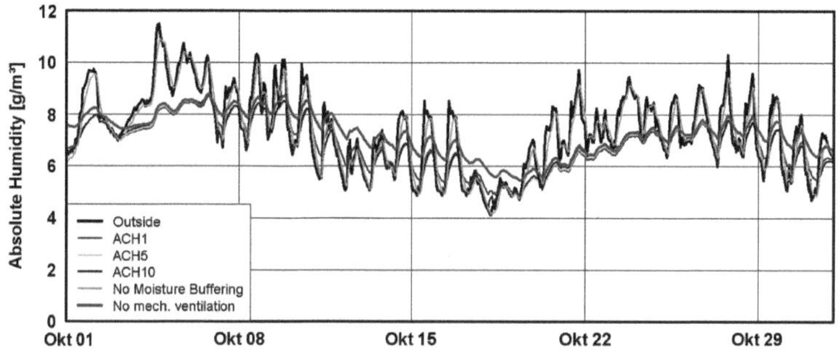

Fig. 6 Absolute humidity outside and inside the church with different parameters for simulation during representative month October

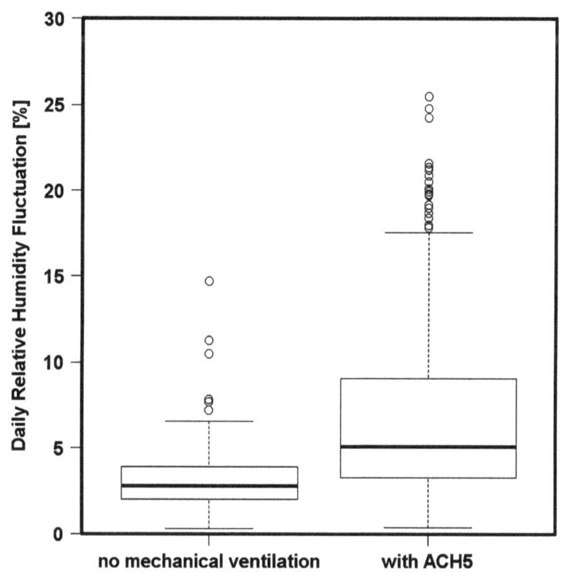

Fig. 7 Boxplots of simulation results for all daily fluctuations of relative humidity in the space without ventilation and the space with ACH 1/h

3.1.5 Discussion

It was found, that daily fluctuations of temperature and relative humidity are higher with mechanical ventilation control compared to the climatically free floating church. Especially in cases with valuable interior artifacts that are sensitive to high RH fluctuations, this may cause problems. The daily fluctuations are compared in Fig. 7 for the case without mechanical ventilation and the ventilation control case with 5 ACH during operation of the mechanical ventilation. The average daily variation is 3 % RH without and 7 % RH with ventilation with maximum daily fluctuations of 15 and 39 % RH respectively.

Fig. 8 Picture of the Gatehall of Lorsch

Table 1 Measured air change rates in the Gatehall of Lorsch (Kießl 1987)

Variations	Air change rate $[h^{-1}]$
Normal condition	0.40
Doors airtight	0.14
Doors and windows airtight	0.06

3.2 Gatehall Lorsch

The Torhalle ("Gate Hall") of Lorsch, also called King's Hall, is one of the best preserved examples of Karoling architecture in Germany, dating back to 760 A.C. The hall is located on the first floor above an entrance gate to a monastery and is believed to have been used as a church originally. It features valuable Romanic wall-paintings. In the 1980s thorough investigations were conducted on the building and its indoor environment which shows high relative humidity all year round Kießl (1987). In 1991 a controlled ventilation system was introduced that promised to lower RH significantly. Already in 1992 it was turned off by a sceptical conservator. Now the building faces the same problems as before requiring new investigations including collecting climate data and evaluating past experience. The aim is to develop a new climatisation strategy for the Gatehall. For this purpose the whole building simulation tool WUFI®Plus is applied, to assess the use of controlled ventilation for the specific location at Lorsch and also to research the possible effects on indoor environmental fluctuations that might be harmful to the ancient wall-paintings. The King's Hall is very well documented regarding its historic building materials and past climate. Also, its special

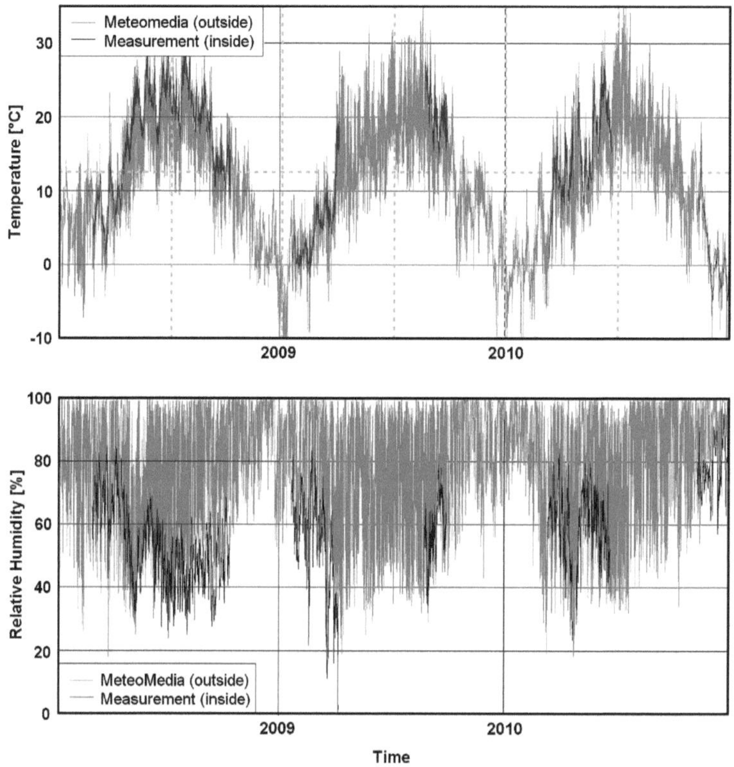

Fig. 9 Measured inner and outer temperature and relative humidity for the years 2008–2010 for the Gatehall Lorsch

construction, sitting above a gate and thus being exposed to exterior climate from all sides without any side effects—like rising damp—makes it especially interesting for whole building simulation. This research is part of the European project Climate for Culture on the effects of climate change on cultural heritage.

The building is a two-story building (see Fig. 8), which has in the lower part a pathway to the monastery church with three big round arches. Above it is the Gatehall, a room with eight small round-arch-windows, which is accessible from both sides trough stairway-towers from the pathway. The Towers are closed down- and upstairs by wooden doors which are probably not airtight. Under the steep roof the room closes with a barrel vaulting, the roof beams are visible. Especially the ancient plaster and fragments of wall-paintings represent the value of the inner surfaces. The facade is designed following the example set by the Romans in a varied "opus reticulatum" technique with coloured natural stone.

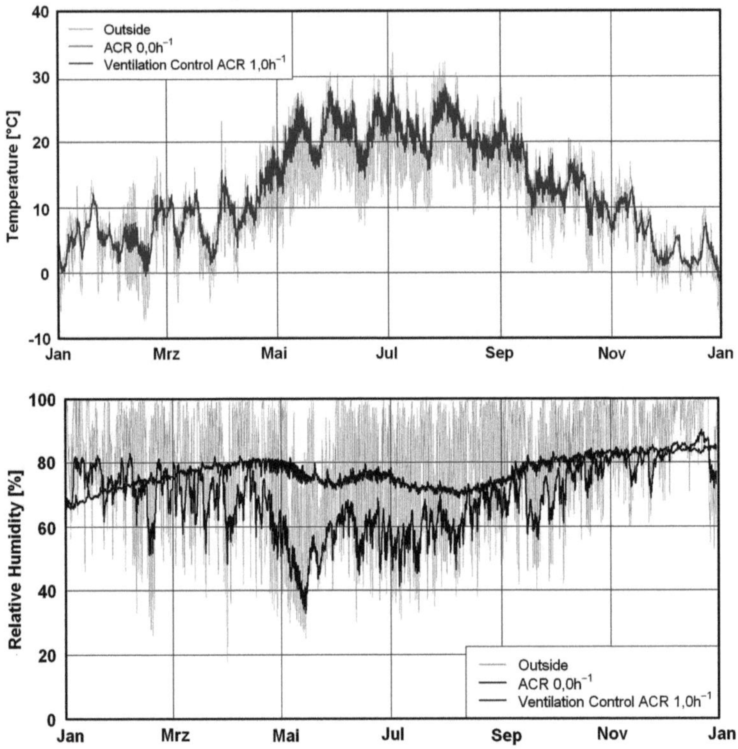

Fig. 10 Simulated inner temperature and relative humidity with an air change rate 0.0 h^{-1} and with a controlled ventilation air change rate 1.0 h^{-1} for the year 2008

3.2.1 Boundary Conditions

Visitors

From the visitor statistics of the last 10 years provided by the Hessische Schlösserverwaltung a generic visitor profile has been specified for the year 2008. Factors for every day of the year and every hour between visiting hours from 10 am to 4 pm were calculated to predict the part of the yearly number of visitors. Depending on the seasonal clothing the generic visitors represent different heat and moisture loads.

Weather Data

On-location the weather was only measured by a simple temperature and humidity data logger, which delivered no satisfying information for use in building

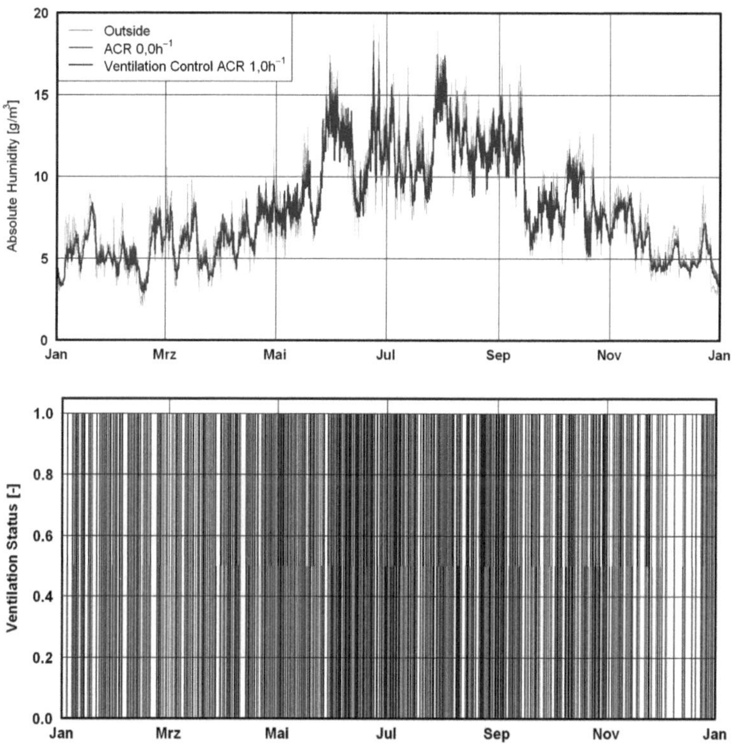

Fig. 11 Simulated absolute humidity and ventilation status with an air change rate 0.0 h^{-1} and with a controlled ventilation air change rate 1.0 h^{-1} for the year 2008

simulation. To guarantee the usability of the weather data from station in Bensheim, 5 km away from Lorsch, both datasets were compared and were almost identical.

Ventilation

The air change rate was measured in a former research project. After an interview with employees of the monastery it became clear, that currently at this time neither controlled ventilation nor any kind of ventilation strategy is established. Windows and doors are opened and closed completely arbitrarily by the employees.

Table 1 shows the results of the measured air change rate in the 1980s. The researchers proceeded in different steps, gradually sealing all openings. The air change rate is low, only 0.4 h^{-1} for normal conditions. It is reduced to 0.26 h^{-1} if only the doors are sealed and to 0.08 h^{-1} if also all windows are sealed.

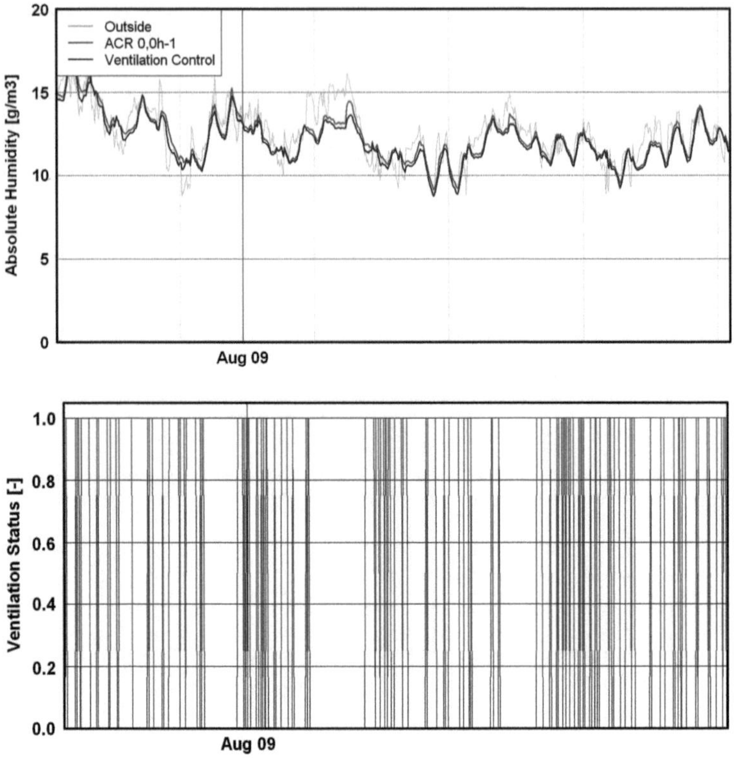

Fig. 12 Simulated absolute humidity and ventilation status with an air change rate 0.0 h^{-1} and with a controlled ventilation air change rate 1.0 h^{-1} for two weeks in August 2008

Outdoor Climate

The grey lines in Fig. 9 show the outdoor temperature and relative humidity in Lorsch. Year 2008 shows a moderate climate, where the outdoor temperature falls rarely below 0 °C and never below −10 °C. The outdoor relative humidity fluctuates considerably, in the summer from <40 % up to 100 % RH. In winter the lower limit is a little bit higher, around 60–80 % RH.

Indoor Climate

The indoor climate was measured over the last 3 years in several shorter time periods. Through the free natural ventilation of the King's Hall the indoor climate follows damped the outdoor climate. In Fig. 9 one can see also, that the indoor temperature follows the upper region of the outdoor temperature, but the indoor relative humidity follows the lower region of the outdoor humidity. Altogether the

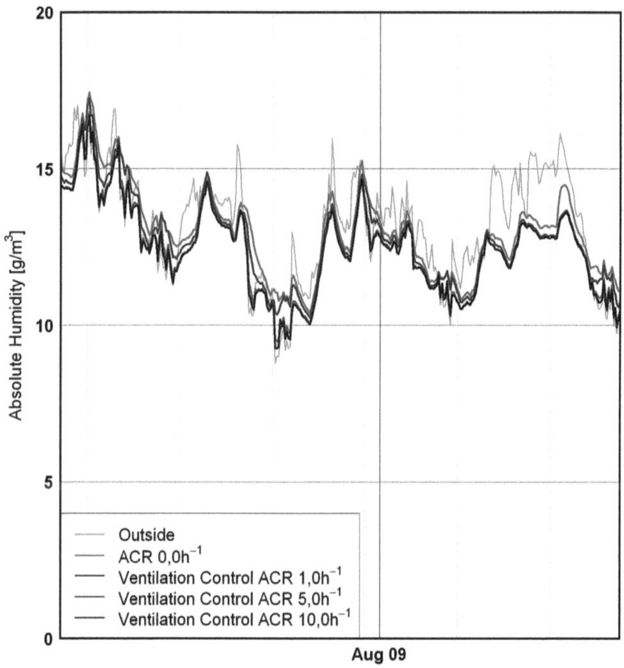

Fig. 13 Simulated absolute humidity for several air change rates for two weeks in August 2008

indoor relative humidity is around 40–80 % RH in summer, but sometimes clearly below. In winter the relative humidity is around 80–100 %.

3.2.2 Parameter Variations

To compare the influence of natural ventilation and controlled ventilation the air change rate was modified in several simulations. A simulation was run with a constant air change rate of 0.0 h^{-1}. Further simulations were run with controlled ventilation and an air change rate 1.0, 5.0 and 10.0 h^{-1}.

3.2.3 Results

In Fig. 10 you can see the difference between an airtight building with an air change rate 0.0 h^{-1} and a building with a controlled ventilation air change rate 1.0 h^{-1}. Even with this low air change rate the level of relative humidity is most of the time lower than without an exchange. The relative humidity nearly fluctuates with the outdoor relative humidity. You can also see, that both the range of the

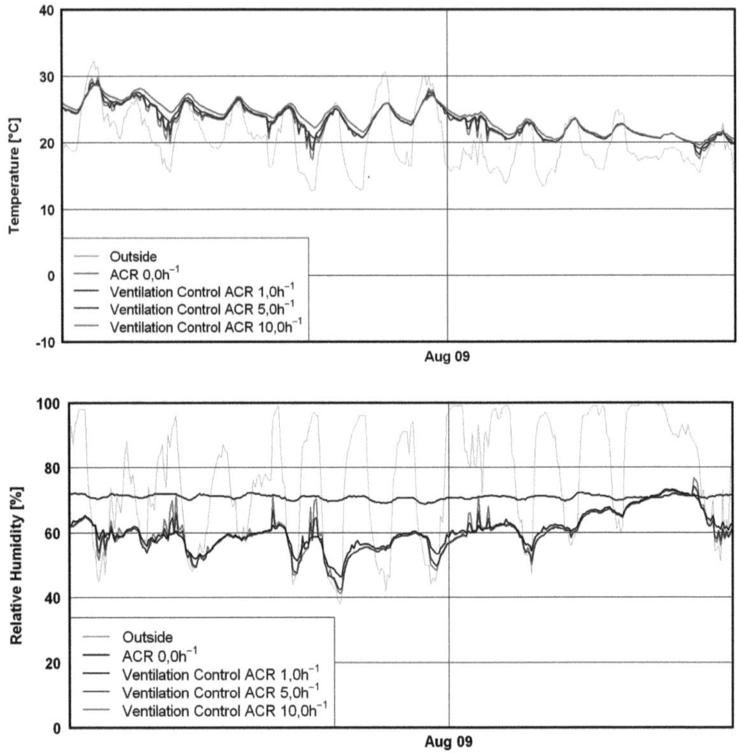

Fig. 14 Simulated inner temperature and relative humidity for several air change rates for 2 weeks in August 2008

daily fluctuations and the seasonal fluctuations of the relative humidity increases with a controlled ventilation.

Figure 11 shows the absolute humidity and the ventilation status for the airtight and the ventilated building. The controlled ventilation can decrease the absolute humidity in the Gatehall by 1–2 g/m^3 with an air change rate of 1.0 h^{-1} over the whole year.

In the detailed Fig. 12 it is obvious that the absolute humidity with controlled ventilation is always below the air tight simulation. But the difference between indoor and outdoor absolute humidity is small. In times when the absolute humidity inside is lower than the outside conditions the ventilation is off.

Higher controlled ventilation air change rate causes lower absolute humidity in the Gatehall. The biggest difference shows the comparison between air change rate 0.0 h^{-1} and air change rate 1.0 h^{-1}, but also between air change rate 5.0 h^{-1} and air change rate 10.0 h^{-1} a difference is still discernible (see Fig. 13).

The level of absolute humidity decreases with higher air change rates for the controlled ventilation. But also the inner temperature decreases. This causes higher fluctuations of the relative humidity (see Fig. 14). The difference can be up to

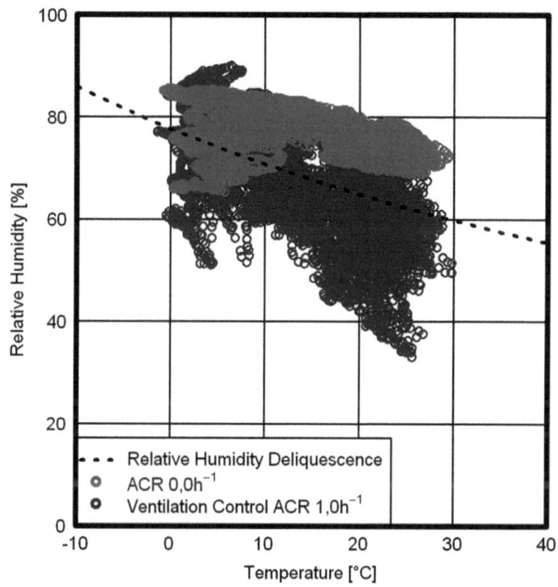

Fig. 15 Simulated inner temperature and relative humidity for an air change rate 0.0 h^{-1} and a controlled ventilation air change rate 1.0 h^{-1} for the year 2008

10 % RH between air change rate 1.0 h^{-1} and air change rate 10.0 h^{-1}. A specific limit of the ventilation has to be found for every situation and building.

3.2.4 Discussion

A high salt load in the plaster and stone of the walls of the Gatehall was found in the research project in the 1980 and 1990s. The salt was examined and it came out, that it was nitrate of ammonium, brought in by pigeon excrements. The deliquescence relative humidity (RHD) can be calculated with a simple model for this salt (Nieveler 1998). Approximately one-third of the measurement points are above the limit. So there is the assumption, that temperature and relative humidity will change inopportunely, that the salt in the walls will get in solution and crystallize several times in a year. This leads into progressive damage like chipping off pieces of plaster or paintings, described in the earlier research projects.

Under the assumption that the surface conditions are close to the temperature and relative humidity conditions inside the room Fig. 15 shows that most of the time steps for the no ventilation case conditions above the RHD are found. In this case the salt in the plaster is in solution. With a ventilation control the range of relative humidity is bigger than without. This could end up in more phase changes of the salt and more damage. In this case the first thoughts of reducing the humidity to save the wall paintings may cause more damage by salt phase changes than it could avoid.

4 Summary and Conclusions

A simulation model of a church and another unheated historic building was implemented. Parameter variations were undertaken, to assess the effectiveness of controlled ventilation to lower humidity conditions inside the buildings. A model of the free floating buildings without mechanical ventilation was compared to the same building with a mechanical ventilation system. This ventilation system was controlled to be active while inner absolute humidity was higher than exterior absolute humidity with the limitation that exterior temperature had to be above 0 °C.

The results of the simulation permitted determining that it is possible to lower the inner absolute humidity during some times of the year with the ventilation control system assessed in this study. The main limitation of the system is that moisture removal can never reduce the inner absolute humidity to a level below exterior conditions. This limits its effectiveness in cases of a free floating building with low inner moisture loads.

Critical conditions due to moisture vented or infiltrated into the building leading to high relative humidity or even liquid water on the envelope surface cannot always be avoided. The daily cycle and other fluctuations in exterior absolute humidity are low. Potential for active ventilation is not always available. Therefore also ventilation systems that create low air change rates allow the full removal of excess moisture. A further increase in air change rate is not necessary. In comparable free floating cases with no inner loads humidity conditions can still be critical. Improvement in the indoor humidity level is bought at the expense of stabile humidity conditions. Fluctuations of temperature and humidity are much higher with controlled ventilation than without. This effect has to be balanced with the small improvements in moisture level. Even if an improvement in terms of moisture level is achievable it is to be assessed if these improvements really reduce possible damages or if it is counterproductive.

A more general understanding of adaptive ventilation as a dehumidification measure in unheated historic buildings is required. As for most of these buildings only low moisture loads prevail, the moisture excess inside the buildings and with it the capacity to remove moisture by controlled ventilation is usually low. Different results can be expected if higher moisture loads, e.g. due to building usage or moisture from the ground, are found. Further research is also required to assess the effects of the interaction of controlled ventilation and heating.

It was shown that it is critical to model coupled heat and moisture transport processes in the envelope and its interaction with the room to achieve realistic simulation results. Relative humidity level and fluctuations can only be represented if moisture buffering effects are taken into account in the simulation.

Acknowledgments This work is part of the European research project "Climate for Culture" (contract nr. 226973) on the impacts of Global Climate Change on Cultural Heritage that is funded by the European Commission.

References

Broström, T., Hagentoft, C.-E., Wessberg, M.: Humidity control in historic buildings through adaptive ventilation—a case study. In: 9th Nordic Symposium on Building Physics, Tampere, Finland (2011)

Holm A. H., Künzel H. M., Sedlbauer K.: The hygrothermal behaviour of rooms: combining thermal building simulation and hygrothermal envelope calculation. In: Eight International IBPSA Conference, Eindhoven, The Netherlands, 11–14 Aug 2003, pp. 499–505 (2003)

Holmberg, J. G.: Environment Control in historical buildings. Royal Institute of technology builiding services engeneering, Bulletin No. 53, Stockholm (2001)

Künzel, H.: Bauphysik und Denkmalpflege. Fraunhofer IRB Verlag, Stuttgart (2010)

Kilian, R.: Statistische Untersuchungen der Klimaschwankungsbreite in unterschiedlich genutzten Kirchenbauten, in WTA Almanach 2007. WTA-Publications, München (2007)

Kießl, K., Reiß, J.: Feuchtetechnische Untersuchungen an Außenwänden der Torhalle Lorsch. IBP-Bericht FtB-1/1986. Holzkirchen (1987)

Kießl, K., Holz, D.: Klimaabhängige Belftungssteuerung und einjährige Feuchtekontrolluntersuchungen im Obergeschoß der Torhalle Lorsch. IBP-Bericht FB-64/1995. Holzkirchen (1995)

Nieveler, U.: Chemisch-analytische Untersuchungen an gasförmigen und partikulären Aerosolbestandteilen im alpinen Bereich. Dissertation, Hamburg (1998)

Sedlbauer, K.: Vorhersage von Schimmelpilzbildung auf und in Bauteilen. Dissertation, Universität Stuttgart (2001)

Degradation Control of Historical Walls with Rising Damp Problems

A. S. Guimarães, J. M. P. Q. Delgado and V. P. de Freitas

Abstract Treating rising damp in the walls of historical buildings is a very complex procedure. Moisture transfer in the walls of old buildings in direct contact with the ground leads to the migration of soluble salts, which are responsible for many building pathologies. This work follows two main lines of research: one theoretical (analytical and numerical) and one experimental. The theoretical part describes an extensive analysis of the phenomenon of rising damp using an analytical equation based on the concepts and methods of unsaturated flow theory, and a numerical validation study. The results show that the simple analytical model clearly describes the rising damp front when compared with the numerical simulations. The influence of wall thickness, boundary conditions, wall composition and material properties such as porosity and sorptivity are analyzed in detail. The experimental part presents the results of the work developed by the Building Physics Laboratory (LFC) to treat rising damp in a historical church, locate in Northern Portugal. The main purpose is to validate the technology of wall base ventilation, for treating rising damp in walls of historic buildings. The analytical model used and the numerical results obtained describe well the observed features of rising damp in walls, verified by in-field tests, who contributed for a simple sizing of the HUMIVENT device to be implement in historic buildings.

A. S. Guimarães (✉) · J. M. P. Q. Delgado · V. P. de Freitas
Faculty of Engineering, LFC—Building Physics Laboratory,
Civil Engineering Department, University of Porto,
4200-465 Porto, Portugal
e-mail: anasofia@fe.up.pt

J. M. P. Q. Delgado
e-mail: jdelgado@fe.up.pt

V. P. de Freitas
e-mail: vpfreita@fe.up.pt

V. P. de Freitas and J. M. P. Q. Delgado (eds.), *Hygrothermal Behavior, Building Pathology and Durability*, Building Pathology and Rehabilitation 1,
DOI: 10.1007/978-3-642-31158-1_6,

Keywords Rising damp · Analytical solution · Numerical results · Historical buildings · HUMIVENT device · Moisture transfer

Nomenclature

a	Cross section area, [m^2]
a_s	Regression constant ($a_s = 0.25$), [–]
A	Water absorption coefficient, [kg/m^{2}s$^{1/2}$]
b	Wall thickness, [m]
b_s	Regression constant ($a_s = 0.50$), [–]
c	Water vapour concentration, [kg/m^3]
c_0	Initial water vapour concentration, [kg/m^3]
c^*	Surface wall water vapour concentration, [kg/m^3]
c_p	Air specific heat ($c_p \cong 1013$), [J/kg °C]
d_r	Variable defined by Eq. (25), [–]
D_m	Molecular diffusion coefficient, [m^2/s]
D_ϕ	Liquid conduction coefficient, [kg/ms]
e	Evaporation potential, [m/s]
e_H	Evaporation potential for the wall base ventilation system, [m/s]
e_s	Evaporation potential for the wall zone with tiles, [m/s]
e_{pm}	Evaporation potential given by Penman–Monteith equation, [m/s]
e_o	Evaporation potential given by Oudin equation, [m/s]
F_c	Capillary force, [N]
F_g	Gravitational force, [N]
F_μ	Viscous force, [N]
g	Gravitational acceleration, [m/s^2]
G	Soil heat flux ($G \cong 0$), [J/m^{2}s]
h	Height, [m]
h_H	Height of the wall base ventilation system, [m]
h_s	Height of the sand, [m]
h_v	Latent heat of phase change, [J/kg]
h_∞	Steady-state height of rise, [m]
i	Volume of liquid absorbed per unit cross section, [m^3/m^2]
J	Number of the day in the year, [–]
L	Length of the saturated wall, [m]
k	Capillary coefficient, [m/s$^{1/2}$]
m	Mass absorbed, [kg]
M	Water molecular weight, [Kg/mol]
n_s	Actual duration of sunshine, [h]
n	Moisture transfer rate, [kg/s]
N	Maximum possible duration of sunshine, [h]
p	Pressure, [Pa]
p_c	Capillary pressure, [Pa]
p_d	Water vapor pressure at dew-point temperature, [kPa]
p_v	Saturation water vapor pressure, [Pa]
Q	Total quantity of water stored within unit length of the wall, [m^3/m]

r	Cylindrical capillary radius, [m]
r_a	Aerodynamic resistance, [s/m]
r_s	Surface resistance ($r_s \cong 70$), [s/m]
Re	Reynolds number ($\mathrm{Re} = \rho_{air}uL/\mu$), [–]
R_a	Extraterrestrial radiation, [J/m^{2}s]
R_g	Ideal gas constant, [J/molK]
R_{ns}	Net solar radiation, [J/m^{2}s]
R_s	Solar radiation, [J/m^{2}s]
S	Sorptivity coefficient ($S = A/\rho_w$), [m/s$^{1/2}$]
Sc	Schmidt number ($Sc = \mu/\rho_{air}D_m$), [–]
t	Time, [s]
T	Temperature, [°C]
T_a	Air temperature, [°C]
T_d	Dew-point temperature, [°C]
u	Capillary rise velocity or Absolute value of air velocity, [m/s]
y	Cartesian co-ordinate, [m]
w	Water content, [kg/m^3]
w_2	Wind speed at a height of 2 m, [m/s]
$w_{80\,\%}$	Water content at 80 % of relative humidity, [kg/m^3]
w_f	Free water saturation, [kg/m^3]
w_s	Variable defined by Eq. (24), [–]
z	Cartesian co-ordinate, [m]
ε	Porosity, [-]
α	Albedo coefficient ($\alpha \cong 0.23$), [–]
δ	Boundary layer thickness, [m]
δ_1	Variable defined by Eq. (26), [–]
δ_p	Vapor permeability, [kg/msPa]
ϕ	Relative humidity,[%]
ϕ	Initial relative humidity, [–]
γ	Liquid surface tension, [N/m]
η	Psychrometric constant ($\eta \cong 0.066$), [kPa/°C]
φ	Angle, [rad]
λ_T	Thermal conductivity, [W/mK]
μ	Liquid viscosity, [Ns/m^2]
μ_a	Air viscosity, [Ns/m^2]
θ	Wetting angle, [Rad]
θ_w	Water bulk volume ($\theta_w = w_f/\rho_w$), [–]
ρ	Density, [kg/m^3]
ρ_{air}	Air density, [kg/m^3]
ρ_w	Water density, [kg/m^3]
Δ	Slope of the vapor pressure curve, [kPa/°C]

1 Introduction

Rising damp, a world-wide phenomenon, is a major cause of decay in masonry materials such as stone, brick and mortar. The conservation of historical buildings has become important nowadays and has developed significantly in recent years. Rising damp in historical buildings may be considered one of the most important of all the different manifestations of damp, leading to the decay of stone materials due to frost/defrost cycles and the presence of salts associated with crystallization (Kunzel 1995) argues that in many cases increased salt content is interpreted as rising damp), decay in elements in contact with the stone walls, such as wooden beams, wooden ceilings and other finishing's, and insanitary conditions caused by excess humidity associated with the development of fungus and mould.

The treatment of rising damp in historical buildings walls is very complex, due to the thickness of the walls and the fact that they are built from different materials. The traditional techniques used to deal with this kind of problem (such as watertight barriers, injection of hydrofuge products, etc.) sometimes prove ineffective or too expensive, justifying the need to find new approaches (see Guimarães et al. 2010).

The Building Physics Laboratory (LFC) at the University of Porto-Faculty of Engineering (FEUP) has been developing important experimental research into the problems of rising damp. In recent years the principles for a treatment technique called the "Wall Base Ventilation System to Treat Rising Damp—HUMIVENT device" has been validated and characterised in experiments. The technique consists of circulating air around the base of very thick walls built from different materials, with a relative humidity saturation distance. Wall base ventilation increases evaporation, which reduces the level of the damp front. This is possible only when the groundwater is lower than the base of the wall (Colombert 1975).

In previous research it was observed that the "wall base ventilation device" reduces the level of the wet front. Following this, a series of experiments were carried out to explore boundary conditions, geometry, engine speed, etc., and it was possible to observe that a hygro-regulated system is essential in controlling possible condensation inside the system. A hygro-regulated system is a mechanical system controlled by probes placed at the inlet and at the outlet. In accordance with pre-programmed criteria, the probes will switch the system on/off.

Historical buildings in Portugal may, in fact, contain walls that are up to 1 m thick, built from various different materials (for example, limestone in the centre, marble in the south and granite in the north). Limestone and granite were used in this study.

This work presents an analytical study of the phenomenon of rising damp (sharp front theory for capillary absorption) developed by Hall and his team (Hall and Hoff 2007; Hall and Hoff 2002; Hall et al. 1984; Hall 1981; Gummerson et al. 1980; Hall 1977), based on the theory of unsaturated flow, together with the numerical results, using WUFI 2D v.3.3, that validate the analytical predictions for steady state height of rise. Although the analytical solution does not provide an exact representation of the physical phenomenon, the solution is very simple and the predictions consistent with practical observations.

Fig. 1 Building pathologies observed in the Portuguese historical church

This work presents the results of our study, which consisted of a large number of numerical simulations on walls composed of different materials (granite and limestone), used to validate the analytical solution developed by Hall and his team. Another important results of this work is the in-field study, which consisted of an analytical analyse related to the use of the wall base ventilation system (HUMIVENT) to treat rising damp phenomena in a historical church, locate in Northern Portugal. The church presented several pathologies (see Fig. 1) as: green spots, fungal developments, surface condensations and damp patches.

The main goal is to estimate the level achieved by the damp front in steady-state. It is very important to have an idea of the height of the rising damp front, h, to predict the improvements of some experimental treatment techniques.

2 The Level Achieved by the Damp Front

2.1 Capillary Absorption

For a vertical cylindrical capillary of radius r with a bottom immersed in liquid the associated capillary pressure, p_c, can be written as

$$p_c = \frac{2\gamma \cos\theta}{r} \tag{1}$$

where γ is the liquid surface tension and θ is the wetting angle between liquid and solid phase. The capillary force, $F_c(= p_c \pi r^2)$, the gravitational force, F_g, and the viscous force, F_μ, given by Hagen–Poiseuille equation (Bird et al. 1960), are given by

$$F_c = 2\pi r\gamma \cos\theta \tag{2}$$

$$F_g = \rho g \pi r^2 h \tag{3}$$

$$F_\mu = 8\pi\mu h \frac{dh}{dt} \tag{4}$$

where ρ is the liquid density, g is the gravity acceleration, h is the meniscus co-ordinate, μ is the liquid viscosity. Using the Newton's law the full dynamic equation obtained is

$$\rho\pi r^2 \left[h\frac{d^2h}{dt^2} + \left(\frac{dh}{dt}\right)^2 \right] = 2\pi r\gamma \cos\theta - \rho g \pi r^2 h - 8\pi\mu h \frac{dh}{dt} \tag{5}$$

Using the Lucas-Washburn approximation, for quasi-steady state rise, low Reynolds number and neglecting the acceleration, the capillary rise velocity, u, is obtained

$$u = \frac{dh}{dt} = \frac{r^2}{8\mu h}\left(\frac{2\gamma\cos\theta}{r} - \rho g h\right) \tag{6}$$

Finally, the analytical solution is given by

$$t = \frac{8\mu}{\rho g r^2}\left(h_\infty \ln\frac{h_\infty}{h_\infty - h(t)} - h(t)\right) \tag{7}$$

with the equilibrium height, h_∞, given by

$$h_\infty = \frac{2\gamma\cos\theta}{\rho g r} \tag{8}$$

and for short-time limit, when $t \to 0$, Eq. (7) results in

$$h(t) = \sqrt{\frac{r\gamma\cos\theta}{2\mu}} t^{1/2} = kt^{1/2} \tag{9}$$

where k is the capillarity coefficient that covers the relationship between the solid and liquid phases, as well as the pore structure inside the solid phase. This result is a generalization of the well-known Washburn equation, describing the horizontal movement of the meniscus in a capillary system, and it can be applied to the vertical movement in building materials in the early stages of capillary rise, neglecting the contribution given by the weight of air columns.

The lack of an adequate method for the determination of liquid height at a given time requires an alternative measurement of the sorptivity, S, which can be calculated according to the relationship:

$$i = St^{1/2} \tag{10}$$

where i is the volume of liquid absorbed per unit cross section in time t (Gummerson et al. 1981). The use of sorptivity as a measure of the capillary absorption properties of porous construction materials is now widespread (Wilson and Hoff 1994), with $k/S = \text{const.}$. As an example, the work developed by Hanžič and Ilic (2003) show that, for water, the ratio between the capillarity coefficient and sorptivity is 5.5 ± 0.6, 5.8 ± 0.6 and 7.1 ± 0.7 in concrete without additives, concrete with an air-entraining agent and concrete with a plasticizer, respectively.

2.2 "Sharp Front" Theory

Analysis of the capillary transport of liquids through porous media is simplified by the use of sharp front models (Green and Ampt 1911; Philip 1969; Hall and Hoff 2002). The rate of advance of the wetting front is found by applying Darcy's Law, $u = -k\nabla p$, to the wetted region, which is assumed to have a uniform and constant water content, θ_w, close to saturation (θ_w is the water bulk volume), in order to obtain the dependence of cumulative absorption on the square root of the elapsed time, $i = St^{1/2}$ (Hall and Hoff 2002). The importance of Darcy's law for unsaturated flow theory is the fact that it describes the advance of the water content profile into an initially dry porous solid. Unsaturated flow theory provides theoretical support for the $t^{1/2}$ law of water absorption: $\Delta m/a = At^{1/2}$, where a is the cross section area of the inflow face and A is the water absorption coefficient (Kunzel 1995). This variable can be related to the quantity S, the sorptivity of unsaturated flow theory, using $S \approx A/\rho_w$, in which ρ_w is the density of water (Gummerson et al. 1980). Furthermore, the sorptivity of a material in relation to a specified test liquid (commonly water) can be easily and accurately measured and is an important material property in building physics (Hall and Tse 1986; Hall 1989; Hall and Hoff 2002).

Considering absorption inflow and evaporation loss to be balanced and disregarding gravitational forces (see Fig. 2), the following differential equation is obtained (Hall and Hoff 2007)

$$\frac{dQ}{dt} = b\frac{S}{2}t^{-1/2} - eh \tag{11}$$

$$\frac{dh}{dt} = \frac{S}{2\theta_w}t^{-1/2} - \frac{eh}{b\theta_w} \tag{12}$$

$$\frac{dh}{dt} = \frac{S^2}{2\theta_w^2}\frac{1}{h} - \frac{eh}{b\theta_w} \tag{13}$$

where Q is the total quantity of water stored within unit length of the wall, h is the damp front height and the total rate of evaporation is given by eh (e is the

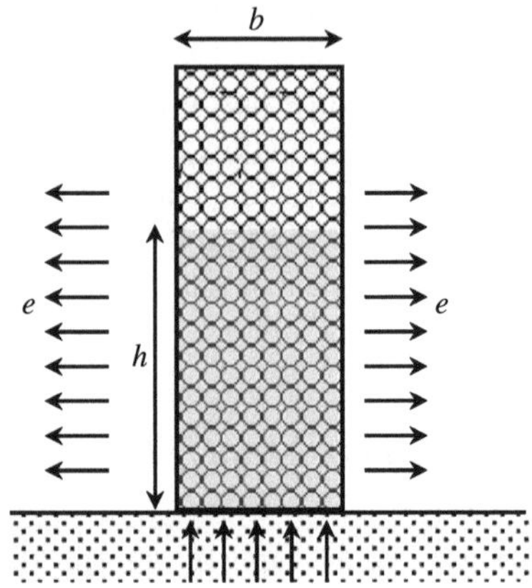

Fig. 2 Sketch of sharp front model

evaporation potential). Considering that $Q = \theta_w bh$ (where b is the wall thickness), the following expression is obtained with $i = St^{1/2} = \theta_w h$, results.

The analytical solution of Eq. (13) is

$$h = S\sqrt{\frac{b}{2e\theta_w}\left[1 - \exp\left(-\frac{2e}{b\theta_w}t\right)\right]} \tag{14}$$

with the equilibrium height ($t \to \infty$), h_∞, given by

$$h_\infty = S\sqrt{\frac{b}{2e\theta_w}} \tag{15}$$

and the sorptivity and water bulk volume given by: $S \approx A/\rho_w$ and $\theta_w \approx w_f/\rho_w$, where w_f is the free water saturation of the building material (see Table 1).

2.3 Evaporation Potential

To estimate the real climate evaporation potential, e, many existing formulae can be applied (Kay and Davies 2008). In this work we compare the performance of Penman–Monteith, e_{pm}, with the simple, temperature-based formula proposed by Oudin et al. (2005), e_o, to evaluate the evaporation potential.

The Penman–Monteith formulation (Monteith 1965) for evaporation potential estimation can be expressed as,

$$e_{pm} = \frac{1}{h_v\rho_w}\frac{\Delta(R_{ns} - G) + \rho_a c_p(p_v - p_d)/r_a}{\Delta + \eta(1 + r_s/r_a)} \tag{16}$$

where h_v is the latent heat flux ($h_v \cong 2.45 \times 10^6$ J/kg), ρ_a is the air density, c_p is the specific heat of air, η is the psychometric constant, R_{ns} is the net solar radiation,

G is the soil heat flux, p_v is the saturation water vapor pressure (kPa), Δ is the slope of the vapor pressure curve, $p_d = e(T_d)$ is the water vapor pressure at dew-point temperature (kPa), r_s is the surface resistance and r_a is the aerodynamic resistance (Allen et al. 1994). Although developed for application with weather data at a daily time-step, mean monthly weather data can be used to compute mean monthly e_{pm}, which will be very similar to the average of daily e_{pm} computed from daily weather data (Allen et al. 1998).

The saturation water vapor pressure, $p_v = p(T_a)$, the slope of the vapor pressure curve, Δ, and the aerodynamic resistance, r_a, are given by

$$p_v = p(T_a) = 0.611 \times \exp\left(\frac{17.27 \times T_a}{T_a + 237.3}\right) \tag{17}$$

$$\Delta = \frac{17.27 \times 237.3 \times p_v}{(T_a + 237.3)^2} \tag{18}$$

$$r_a = \frac{208}{w_2} \tag{19}$$

where T_a is the air temperature, T_d is the dew-point temperature and w_2 is the wind speed at a height of 2 m. The net solar radiation, R_{ns} (in MJ/m^{2}d), resulting from the balance between incoming and reflected solar radiation is given by:

$$R_{ns} = (1 - \alpha) \times R_s \tag{20}$$

where α is the Albedo reflection coefficient, which is 0.23 for the hypothetical grass reference crop. If the solar radiation, R_s (in MJ/m^{2}d), is not measured, it can be calculated with the Angstrom formula, which relates solar radiation to extra-terrestrial radiation and relative sunshine duration

$$R_s = \left(a_s + b_s \frac{n_s}{N}\right) \times R_a \tag{21}$$

with

$$N = \frac{24}{\pi} w_s \tag{22}$$

where n_s is the actual duration of sunshine, N is the maximum possible duration of sunshine or daylight hours, R_a is the extraterrestrial radiation (in MJ/m^{2}d), a_s is a regression constant, expressing the fraction of extraterrestrial radiation reaching the earth on overcast days ($n = 0$) and $a_s + b_s$ is the fraction of extraterrestrial radiation reaching the earth on clear days ($n_s = N$). Where no actual solar radiation data are available and no calibration has been carried out for improved a_s and b_s parameters, the values $a_s = 0.25$ and $b_s = 0.50$ are recommended.

The simple temperature-based evaporation potential formulation suggested by Oudin et al. (2005), based on a study of the performance of over 25 existing evaporation potential formulations, is given by

$$e_o = \begin{cases} \frac{R_a}{h_v \rho_w} \frac{T_a + 5}{100} & \text{if } T_a + 5 > 0 \\ 0 \text{ otherwise} \end{cases} \tag{23}$$

where R_a is extraterrestrial radiation, which is dependant on latitude and Julian day only (Allen et al. 1994). The extraterrestrial radiation, R_a (in MJ/m^{2}d), for each day of the year and for different latitudes can be estimated from the solar constant, the solar declination and the time of the year by

$$R_a = \frac{24 \times 60}{\pi} \times 0.082 \times d_r \times [w_s \sin(\varphi) \sin(\delta_1) + \cos(\varphi) \cos(\delta) \sin(w_s)] \tag{24}$$

with

$$d_r = 1 + 0.033 \times \cos\left(\frac{2\pi}{365} J\right) \tag{25}$$

$$\delta_1 = 0.409 \times \sin\left(\frac{2\pi}{365} J - 1.39\right) \tag{26}$$

$$w_s = \cos^{-1}[-\tan(\varphi) \times \tan(\delta)] \tag{27}$$

$$\varphi = \frac{\pi}{180} \times [\text{latitude in decimal degrees}] \tag{28}$$

where J is the number of the day in the year between 1 (1 of January) and 365 or 366.

Finally, to estimate the evaporation potential we consider that Porto latitude is 40.15 °. So using Eq. (24) the average extraterrestrial radiation for each day of the year is 321.9 J/m^{2}s, resulting in a evaporation potential, e, of 0.0017 min/min and 0.0016 min/min, with Eqs. (16) and (23), respectively.

For the climate inside the laboratory, the evaporation potential, considering a constant temperature and a relative humidity, 20 °C and 60 %, respectively, and an air velocity inside the laboratory of approximately 0.04 m/s, was given by:

$$e = \frac{n}{\rho_w} = \frac{1}{\rho_w} \frac{D_m}{\delta} (c^* - c_0) \tag{29}$$

where n is the moisture transfer rate, D_m is the water vapor diffusivity in air, ρ_w is the water density, δ is the boundary layer thickness, c^* is the surface wall water vapor concentration and c_0 is the initial water vapor concentration. For laminar boundary layers over a flat plate, the Blasius solution gives (Schlichting 2004):

$$\delta \approx 4.91 \sqrt{\frac{\mu_a . h}{\rho_a . u}} \tag{30}$$

where ρ_a is the air density ($\rho_a = 1.204\,\mathrm{kg/m^3}$ at $T_a = 20°C$), μ_a is the air viscosity ($\mu_a = 1.82 \times 10^{-5}\,\mathrm{kg/ms}$), u is the air velocity ($u \approx 0.04\,\mathrm{m/s}$) and h is the height of the wall analyzed numerically and analytically ($h = 1.30\,\mathrm{m}$).

The concentration gradient, $c^* - c_0$ (kg/m³), was given by:

$$c^* - c_0 = \frac{p_v.M(1-\phi_0)}{R_g(273.15+T_a)} = 6.03 \times 10^{-3} \exp\left(\frac{17.08\,T_a}{273.15+T_a}\right)\frac{M(1-\phi_0)}{R_g(273.15+T_a)} \tag{31}$$

where M is the water molecular weight ($M = 0.018\,\mathrm{kg//mol}$), R_g is the ideal gas constant ($R_g = 8.205 \times 10^{-4}\,\mathrm{m^3atm/molK}$), p_v is the saturation water vapor pressure and ϕ_0 is the initial relative humidity ($\phi_0 = 0.6$).

Finally, the following expression for the water vapor diffusivity in air was used (Kimball et al. 1976):

$$D_m = 2.23 \times 10^{-5}\left(\frac{273.15+T_a}{273.15}\right)^{1.5} \tag{32}$$

resulting, in a evaporation potential value of 7.3×10^{-5} mm/min, for the laboratory climate.

2.4 The Wall Base Ventilation System

The "HUMIVENT" technique consists of ventilating the base of walls using a natural ventilation process or by installing a hygro-regulated mechanical ventilation system (see Fig. 3). In terms of analysis, we considered a wall base ventilation system along which air is flowing close to the "saturated" wall ($0 < z < L$).

In laminar flow, the transport of moisture and temperature across the boundary layers is controlled by molecular diffusion. The Reynolds number, Re, may be interpreted as the ratio of the flow of destabilising forces to stabilising forces (viscosity). Stable laminar flow is thus characterized by low Re values. Laminar flow occurs at Reynolds numbers of less than approximately 3×10^5. In our study $\mathrm{Re} < 10^5$, in all experiments.

If the analysis is restricted to situations in which the moisture transfer boundary layer is thin and if a small control volume is considered inside this boundary layer, with side lengths δz, δy and unity (perpendicular to the plane of the figure), a steady state material balance in the solute leads to

$$u\frac{\partial c}{\partial z} = D_m\frac{\partial^2 c}{\partial y^2} + D_m\frac{\partial^2 c}{\partial z^2} \tag{33}$$

where D_m is the molecular diffusion coefficient, in the cross stream and in the stream wise directions. If the boundary layer is thin compared to the length of the saturated wall, the last term on the right-hand side of Eq. (33) is likely to be negligible. For the situation outlined the boundary conditions are,

$$c = c_0 \quad z = 0 \quad y > 0 \tag{34a}$$

$$c = c^* \quad z > 0 \quad y = 0 \tag{34b}$$

$$c \to c_0 \quad z > 0 \quad y \to \infty \tag{34c}$$

and the analytical solution is given by (Crank 1975)

$$\frac{c - c_0}{c^* - c_0} = \operatorname{erfc}\left(\frac{y}{2\sqrt{D_m z/u}}\right) \tag{35}$$

where c_0 is the bulk concentration of water vapour, c^* is the surface wall concentration of water vapour, and u is the air velocity. The flux of evaporation, N, at the saturated wall surface may be obtained from (35) as

$$N = -D_m\left(\frac{\partial c}{\partial y}\right)_{y=0} = (c * - c_0)\left(\frac{D_\mathrm{m}}{\pi\, z/u}\right)^{1/2} \tag{36}$$

Finally, it is possible to obtain the total rate of solid evaporation, n, over the whole wall surface

$$n = \int_0^L N\, h_H\, dx = (c^* - c_0)\, h_\mathrm{H}\, L\left(\frac{4D_m}{\pi L/u}\right)^{1/2} \tag{37}$$

where L is the length of the saturated wall and h_H is the height of HUMIVENT system.

The configurations tested (boundary conditions) and used in the numerical analysis are shown in Fig. 3 (walls with a thickness of 200 or 500 mm and a height of 1,500 mm). The only difference between Configurations 1 and 2 was that one side of the wall in Configuration 2 was completely covered with tiles. The aim of using these two boundary conditions was to analyze the influence of waterproof materials (such as tiles). In Configuration 3 a ventilation box was placed on both sides of the wall (see Fig. 3), with an area of (h_H = 200 mm) and sand, on the both sides to a height of 450 mm (h_s = 250 mm). The sand was saturated (100 % RH) during the tests.

With regard to the rising damp steady state for Configuration 3 (with saturated sand placed on both sides of the wall up to a height of 450 mm), Eq. (15) produces

If we consider a rising damp steady state, involving a wall buried in sand (with height h_s and evaporation potential e_s) and a wall base ventilation system (with height h_H and evaporation potential e_H) located on both sides above the base, Eq. (15) produces

$$\frac{bS^2}{2\theta_w h_\infty} = e(h_\infty - h_s - h_H) + e_H h_H + e_s h_s \tag{38}$$

with e_H (related to HUMIVENT) given by (Guimarães et al. 2010)

$$e_H = \frac{n}{L.h_H \rho_w} = \frac{(c^* - c_0)}{\rho_w} \left(\frac{4D_m}{\pi L/u}\right)^{1/2} \tag{39}$$

where ρ_w is the water density. The steady-state height of rise given by

$$\begin{aligned} h_\infty &= \left(1 - \frac{e_s}{e}\right)\frac{h_s}{2} + \left(1 - \frac{e_H}{e}\right)\frac{h_H}{2} + \\ &+ \sqrt{\frac{bS^2}{2\theta_w e} + \left(\frac{e_s}{e} - 1\right)^2 \frac{h_s^2}{4} + \left(\frac{e_H}{e} - 1\right)^2 \frac{h_H^2}{4} + \left(1 - \frac{e_s}{e} - \frac{e_H}{e} + \frac{e_s e_H}{e^2}\right)\frac{h_s h_H}{2}}. \end{aligned} \tag{40}$$

Noting that the evaporation potential related with sand placed on both sides of the wall above its base, e_s, will be negligible (sand water saturated) in comparison with the other evaporation rates, e and e_H, and Eq. (40) reduces then to

$$h_\infty = \frac{h_s}{2} + \left(1 - \frac{e_H}{e}\right)\frac{h_H}{2} + \sqrt{\frac{bS^2}{2\theta_w e} + \frac{h_s^2}{4} + \left(\frac{e_H}{e} - 1\right)^2 \frac{h_H^2}{4} + \left(1 - \frac{e_H}{e}\right)\frac{h_s h_H}{2}} \tag{41}$$

For the case of $h_H = 0$, and $e_s = 0$ (Configurations 1 and 2 of Fig. 3), Eq. (41) reduces to

$$h_\infty = \frac{h_s}{2} + \sqrt{\frac{bS^2}{2e\theta_w} + \frac{h_s^2}{4}} \tag{42}$$

3 Numerical Simulation

Simulation programs that evaluate changes in the moisture content and temperature inside walls are essential tools for simulating the behavior of walls in the presence of moisture dependent on internal and external climatic conditions (Bomberg 1974).

The calculation program used in the numerical simulations for the experimental and analytical validations was WUFI-2D v.3.3, developed by the Fraunhofer Institute for Building Physics. The governing equation for moisture transport is (Holm and Kunzel 2003; Kalamees et al. 2006)

$$\frac{dw}{d\phi}\frac{\partial \phi}{\partial t} = \nabla(D_\varphi \nabla \phi + \delta_p \nabla(\phi p_a)) \tag{43}$$

Table 1 Material properties (Torres and de Freitas 2007)

		Limestone		Granite
Bulk density, ρ (kg/m^3)		2,155		2,550
Heat capacity, c_p (J/kgK)		1,000		850
Porosity, ε (%)		19.7		12.5
Thermal conductivity, λ_T (W/mK)		1.33		2.30
Vapor diffusion resistance factor, μ (–)	Dry cup:	41		55
	Wet cup:	29		
Moisture storage function, w (kg/m^3)	ϕ (%)	*Adsor*	*Desadsor*	
	4.0	0.521	0.951	Generate by Eq. (45)
	11.2	0.593	1.150	
	34.8	0.872	1.239	
	58.6	1.043	1.628	
	76.3	1.237	2.360	
	80.0	1.334	—	
	84.2	1.584	—	
	92.1	2.831	4.144	
Capillary transport coefficient, D_w (m^2/s)	$w = 1.7$ kg/m^3	6.6×10^{-11}		Generate by Eq. (47)
	$w = 188$ kg/m^3	6.2×10^{-8}		
Water absorption coefficient, A (kg/m^{2}s$^{1/2}$)		0.024		0.012
Free-water saturation, w_f (kg/m^3)		177		112

and the governing equation for heat transport is

$$\frac{dH}{dT}\frac{\partial T}{\partial t} = \nabla(\lambda_T \nabla T) + h_v \nabla(\delta_p \nabla(p_a)) \tag{44}$$

where dH/dT is the heat storage capacity of the moist building material, $dw/d\phi$ is the moisture storage capacity, λ_T is the thermal conductivity, D_ϕ is the liquid conduction coefficient, δ_p is the water vapor permeability, h_v is the evaporation enthalpy of the water, p_a is the water vapor saturation pressure, T is the temperature and ϕ is the relative humidity (Evrard and De Herde 2010).

In the numerical analysis used to validate the analytical study only the change in relative humidity was considered important, since the experiments took place under isothermal conditions.

Hygrothermal modeling offers a powerful tool for predicting heat and moisture transport through multi-layer building assemblies. In this work, the hygrothermal model was used to compare the results of three configurations (see Fig. 3), with two type of monolithic walls, with different wall thicknesses and under different natural conditions. The simulations used two different climates, the real climatic variables of Porto and a laboratory climate, and actual material properties to determine the height of the rising damp front (see Table 1).

The numerical simulations, with real climate, were made using climate data for Porto provided by a software tool (METEOTEST 2008). This program calculates the hourly values for all parameters using a stochastic model and the resulting weather data files are produced in a variety of formats. The weather data used as

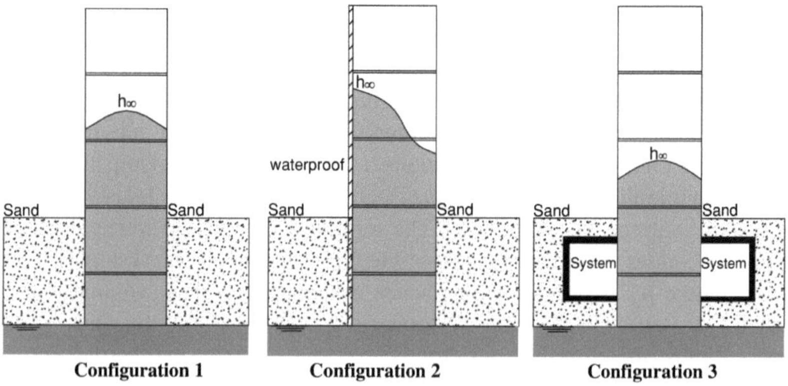

Fig. 3 Different boundary conditions studied

input for the model was temperature (°C), relative humidity (–), precipitation (mm/h) and global solar radiation on a horizontal surface (W/m^2). The short wave radiation absorptivity and long-wave radiation emissivity considered were 0.4 (stucco-normal bright) and 0.9, respectively, and the initial conditions within the element were $\phi = 80\ \%$ and $T = 20$ °C.

The other numerical simulations used the climatic conditions inside the laboratory. The temperature and a relative humidity were considered constant, 20 °C and 60 %, respectively.

The vapor diffusion thickness value used was zero (no coating) and the interior heat transfer coefficient was constant and equal to 8 W/m^{2}K. The exterior heat transfer coefficient only contained the convective element and was considered independent of wind (at a constant value of 17 W/m^{2}K).

The simulation program provided the water content, relative humidity and temperature at all the points on the pre-defined grid and their evolution during the simulation. It is important to bear in mind that in the numerical simulations with granite the following approximation for moisture storage function was used (Kunzel 1995)

$$w(\varphi) = w_f \frac{(b-1)\phi}{b-\phi} \tag{45}$$

with

$$b = 0.8 \frac{w_{80\ \%} - w_f}{w_{80\ \%} - 0.8 w_f} \tag{46}$$

and the capillary transport coefficient for the suction process is estimated by

$$D_{\mathrm{ws}} = 3.8 \left(\frac{A}{w_f}\right)^2 1000^{(w/w_f - 1)} \tag{47}$$

4 Numerical and Analytical Results

Table 2 shows the analytical results (using Eqs. (41) and (42)) obtained for the three different boundary conditions (configurations) in Laboratory interior conditions and Porto climate. For configurations 1 and 3 the rising damp front was obtained in the middle of the wall ($x = b/2$), and in configuration 2 this value was obtained in the waterproof side ($x = 0$). The sorptivity values used were calculated using the A values (water absorption coefficient) shown in Table 1, with $S \approx A/\rho_w$, where ρ_w is the density of water, and the water bulk volume, θ_w, is given by $\theta_w = w_f/\rho_w$. Configuration 2 shows evaporation on only one side of the wall and $e/2$ must therefore be considered.

The results presented in Table 2 show that the simple analytical model proposed by Hall et al. (1984), clearly describes the rising damp front when compared with the numerical simulations, which provide a qualitative validation of Eqs. (41) and (42).

The results, also, show that with a controlled climate (in laboratory) the similarities between the numerical and analytical results were better (with a relative error of approximately 8.3 %, in average) when compared with the results obtained with a real climate (with a relative error of approximately 19.6 %, in average). This is an expectable result considering the influence of several climate variables in the numerical simulation, as radiation, precipitation, temperature, wind velocity and direction, throughout the year, and taking into account that in the analytical analysis, (Eqs. 41 and 42) the evaporation potential considered was an annual average.

Figure 4 show an example of the variation in the rising damp front in terms of the relative humidity and water content for different boundary conditions, wall composition and thicknesses and climates. The graphical representation shows an infinitesimal section of the wall thickness in the center of the wall.

Table 2 shows that rising damp increases when drying conditions are hampered by extra waterproof coatings (see Configurations 1 and 2). Coatings with reduced porosity cannot be used on walls without renderings and hiding anomalies behind a new built wall would reduce space, and hide original walls. In addition, the presence of a wall base ventilation system on both sides prevents the damp front (Configuration 3).

Table 2 also show that the steady-state heights of rise, h_∞, obtained with the Laboratory conditions were highest, when compared with the results obtained with the Porto climate, in accordance with the nonexistence of solar radiation and lower values of evaporation potential in the Laboratory.

Related to the material properties, limestone presented a higher capillary ascension and granite a lower capillary ascension. This difference is explained by the fact that rising damp increases in materials with higher sorptivity values ($S \approx A/\rho_w$) and porosity values (water bulk volume is a function of porosity), as shown in Eq. (15).

Table 2 Comparison of the numerical results (obtained after a one-year simulation, with an error range of less than 5 %) with the analytical solution given by Eq. (41) or (42). **a** In Laboratory and **b** in Porto

Laboratory conditions	Configuration 1				Configuration 2				Configuration 3			
h_∞ (mm)	b = 200 mm		b = 500 mm		b = 200 mm		b = 500 mm		b = 200 mm		b = 500 mm	
	Lime stone	Granite	Lime stone	Granite	Lime stone	Granite	Lime stone	Granite	Lime stone	Granite	Lime stone	Granite
Eq. (41)–(42)	784	618	1065	781	983	733	1391	979	652	470	944	649
WUFI 2D	790	542	1073	738	1025	655	1460	902	630	369	908	491
Relat. Error	0.8 %	12.2 %	0.8 %	5.5 %	4.3 %	10.6 %	4.9 %	7.8 %	3.4 %	21.5 %	3.8 %	24.3 %
Porto climate	b = 200 mm		b = 500 mm		b = 200 mm		b = 500 mm		b = 200 mm		b = 500 mm	
h_∞ (mm)	Lime stone	Granite	Lime stone	Granite	Lime stone	Granite	Lime stone	Granite	Lime stone	Granite	Lime stone	Granite
Eq. (41)–(42)	476	460	510	475	499	471	559	498	467	451	502	467
WUFI 2D	638	466	654	599	612	545	645	552	490	235	630	454
Relat. Error	34.1 %	1.2 %	28.3 %	26.0 %	22.7 %	15.8 %	15.3 %	10.8 %	5.0 %	47.9 %	25.6 %	2.7 %

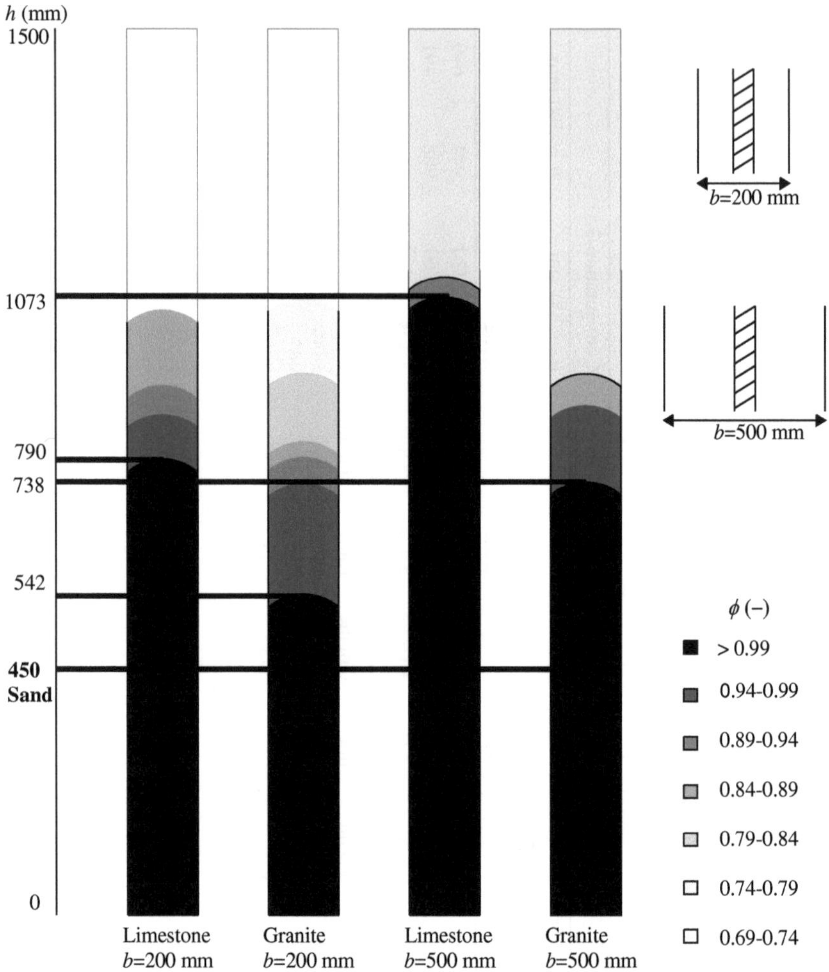

Fig. 4 Variation in rising damp front with relative humidity (Configuration 1), for limestone and granite, with different wall thicknesses, in Laboratory interior conditions

The numerical results, presented in Table 2, show that the rising damp front increases with the wall thickness, *b*. Torres and Freitas (2007) obtained similar results in a previous experimentally and numerically work. To analyze the relation between the rising damp front height and the wall thickness, several numerical and analytical results of a simple configuration were obtained and presented in Table 3. The results show that this increases is approximately a function of the square-root of the wall thickness, with exponents between 0.45 and 0.54 (see Fig. 5). This is an expectable result if we consider the sharp front theory developed by Hall and Hoff (2002) and Eq. (15). This simplified analytical model analyzes the capillary transport of liquids through porous media by the balance between absorption inflow and evaporation loss.

Table 3 The wall thickness influence in rising damp phenomenon (numerical and analytical results)

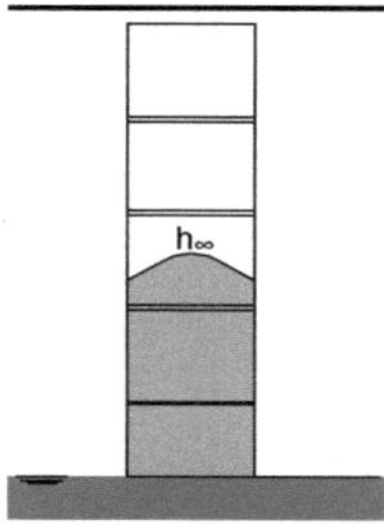

h_∞ (mm)	Laboratory climate							
	Limestone				Granite			
b (mm)	200	400	600	800	200	400	600	800
Eq. (15)	512	724	886	1024	322	455	557	643
WUFI 2D	553	752	924	1102	259	402	477	546
Relat. Error	8.1 %	3.9 %	4.2 %	7.7 %	19.5 %	11.6 %	14.4 %	15.1 %
h_∞ (mm)	Porto climate							
Eq. (15)	111	156	191	221	69	98	120	139
WUFI 2D	143	178	247	254	93	124	156	178
Relat. Error	29.4 %	13.9 %	29.0 %	14.9 %	33.9 %	26.2 %	29.6 %	28.1 %

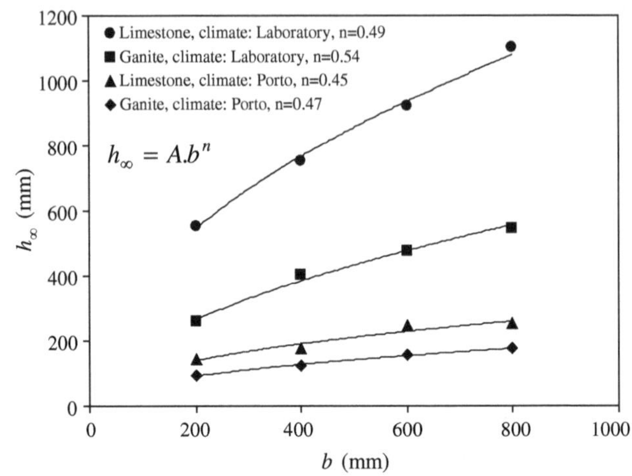

Fig. 5 The wall thickness influence in rising damp phenomenon

5 Experimental Study

The hygro-regulable system was installed in a church in Northern Portugal (see Fig. 6). Inside the building, two hygro-regulable mechanical ventilation subsystems were installed (see Fig. 7). In the Southside subsystem, air was admitted through grids located inside the building, and was extracted into the cloister. Extraction was controlled by a hygro-regulable engine of variable speed (Freitas and Guimarães 2008). The system had two probes for measuring relative humidity and temperature, two transmitters, a control module and a data acquisition system for recording results (see Fig. 8).

The system installed in the church initially began operating whenever the relative humidity at the exit was 5 % higher than the relative humidity at the entrance. The idea was to admit dry air comparing to the air inside the system. This criterion was found to be inadequate, as it meant that the system was operating at periods when condensation occurred inside it. Consequently, a new criterion was proposed with a view to optimizing the system, based upon the difference in vapour pressure (ΔP) at the exit and entrance. The system now began functioning whenever the ΔP was positive.

6 Interpretation of Experimental Results

Figures 9 and 10 show, respectively, the relative humidity and temperature differences, at the inlet and at the outlet of the subsystems. These values were directly read from the data-logger installed.

Fig. 6 Historical church analysed in Northern Portugal

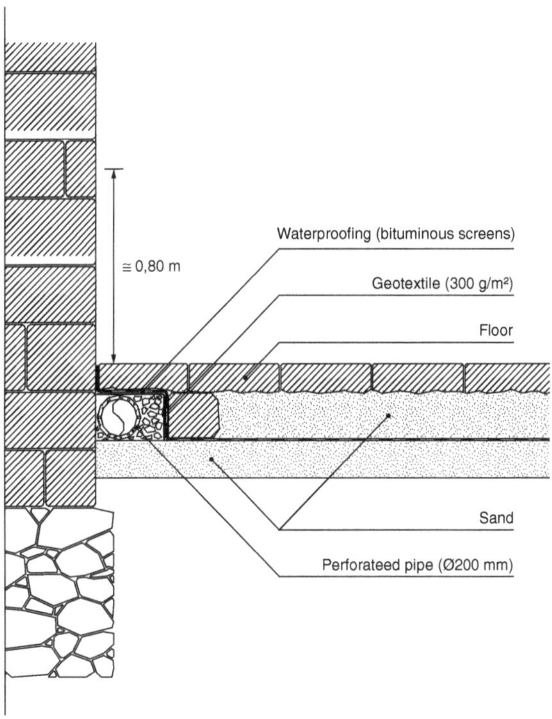

Fig. 7 The wall base ventilation system used in the historical church

These figures show that the relative humidity at the entrance of these subsystems displays high values throughout the year, with an average value close to 80 % at South and close to 85 % (sometimes 100 %) at North. During summer months the average inlet temperature was higher than the temperature at the exit. On the other hand, during the winter months the average inlet temperature was lower than the temperature at the exit. The values of the temperature inside the system, next to

Fig. 8 Data acquisition and recording system

the outlet, followed up on the values of the air temperature at the entry, although with a clear gap, especially on South subsystem.

Figures 11a and b shows the vapour pressure differences between the inlet and outlet subsystems, and the daily average values to better assess their behaviour, in particular, the possibility of condensations in the interior of the subsystems.

Analyzing the results of the study developed "in field" it can be concluded the following:
For north subsystem:

- The relative humidity on entry remained high throughout the year with average values of about 85 %, which in conjunction with the high outside temperature in the summer, contributed to the occurrence of condensation which may occur within the system during these months, as stated to the subsystem. However, as the air is admitted from the outside, the variation of vapour pressure at the entry is very significant, although still in the summer, the wall can be drying. Thus, despite condensations occur during the summer months, there are also periods of time where the system works effectively.
- In winter, the exterior relative humidity presents values of 95 %, although the temperature outside be often lower than the temperature inside the system, condensations occur within the system once the air admitted is already very close to the saturation.
- Temperature values inside the ventilation system, next to the output, followed up on the values of the air temperature at the entry, although with a slight time lag. The thermal inertia is an important factor and that justifies this behaviour.
- The ventilation device was connected over 52 % of the time, according to the second criteria for the operation referred to. In the remaining 48 % of the time the fan was turned off, where possibly condensations occurred within the system. The time periods in which the system was shut down occurred during the summer and during the winter.
- During the summer the subsystem north withdrew more water than during the winter months.

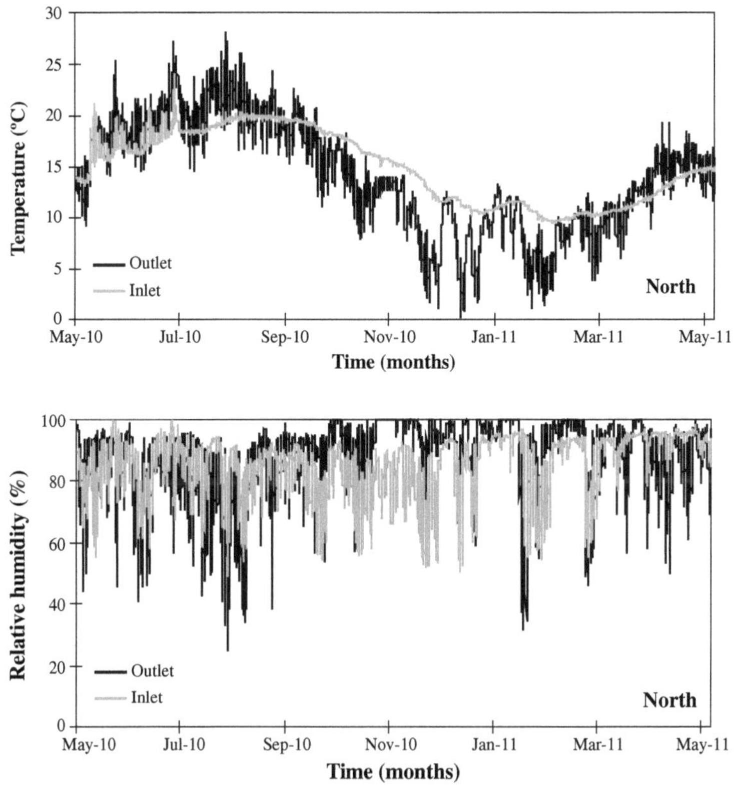

Fig. 9 Temperature and relative humidity variation of northern subsystem

For south subsystem:

- The relative humidity on entry remained high throughout the year with average values of around 80 %, which in conjunction with high temperatures of air admitted in the summer, contributed to the occurrence of condensation which may occur within the system during these months. In summer the system in its interior, presents lower temperatures on air admitted, and a high relative humidity. When the hot and humid air admitted, is within the system, can easily occur condensations.
- In the winter, despite the air relative humidity admitted stay in 80 %, the temperature at the inlet becomes lower than the temperature on the interior of the system. Thus, the damp and cold air, when enter in the system heats up and "dries", allowing a good behaviour of this subsystem during winter months.
- Temperature values inside the ventilation system, at the outlet, followed up on the values of the air temperature at the entry though with a clear gap. The thermal inertia is an important factor that justifies this behaviour.
- The ventilation device was connected over 82 % of the time, according to the second criteria for the operation referred to. In the remaining 18 % of the time

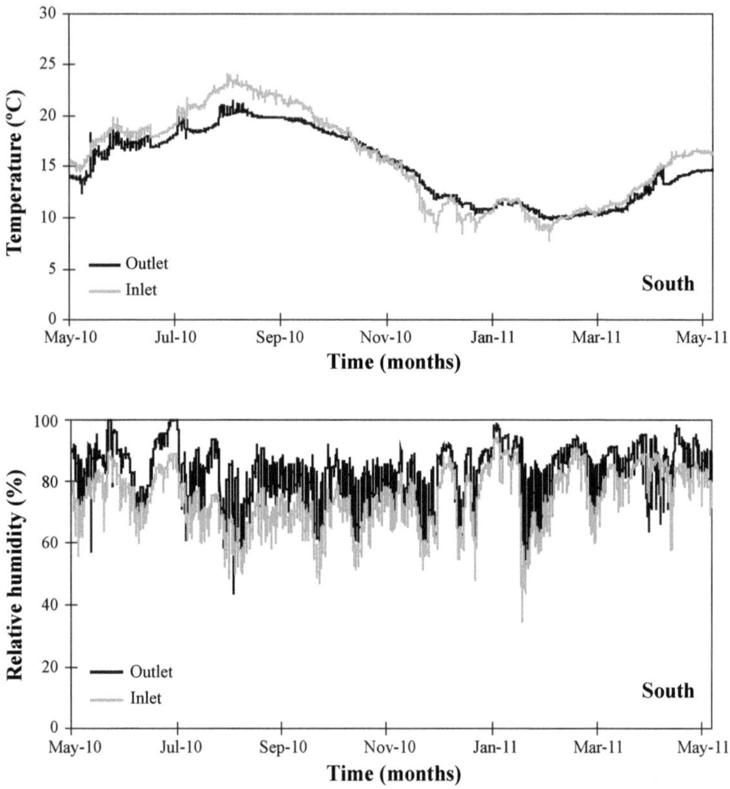

Fig. 10 Temperature and relative humidity variation of southern subsystem

the fan was turned off, where possibly condensations occurred within the system. The time periods in which the system was shut down occurred during the summer months.

- During the winter a greater amount water of subsystem south withdrew compared with the summer months, since in these months occurred condensations within the system, several times. During the summer period the system crashed at various times, reducing the amount of water withdrawn in this period.

Finally, Figs. 12a and b shows the functioning periods of the ventilator, in a north and south wall, and the importance of HUMIVENT device in rising damp treatment, by the quantity of water vapour extracted during the five months analyzed (454.6 kg in the north wall and 386.9 kg in the south wall). These figures show that the operation periods of the subsystem South was majority during the winter months with longer periods of stopping during the summer months. On the other hand, the North subsystem presents a high number of interruptions distributed throughout the year. In this subsystem the air was admitted from the outside of the Church and, in consequence the vapour pressure values were less stable

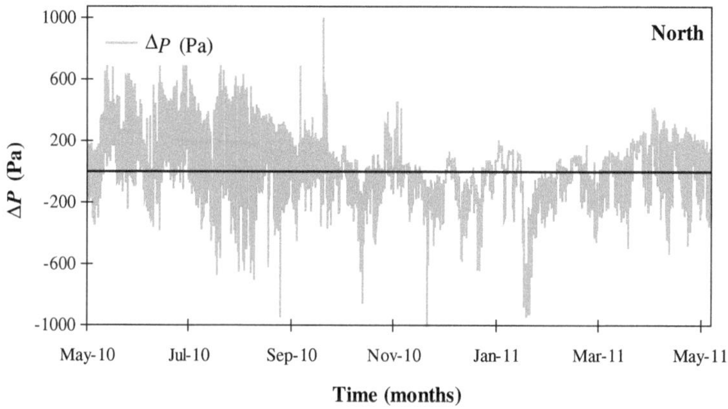

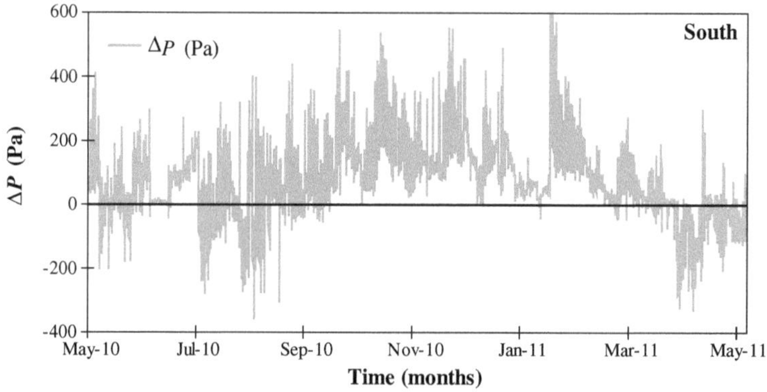

Fig. 11 Vapour pressure differential and functioning of system. **a** North wall and **b** South wall

(sometimes with very abrupt variations), which also could justify the high number of fan interruptions.

The research group is currently studying whether these values may have adverse consequences in terms of the long-term deterioration of the wall materials (Watt and Colston 2000). The entry of air with very low relative humidity could generate the crystallization of salts existing in the building materials, threatening its durability. For this reason, the relative humidity value at the entrance had to be limited. The relative humidity scores recorded, which range from 60 to 95 %, are not considered to present a risk of salt crystallization/dissolution inside the system, consider the salts detected in those area. However, the problem might arise in another type of external climate or with other salts type. In that case the HUMI-VENT device must also limit the inlet relative humidity value.

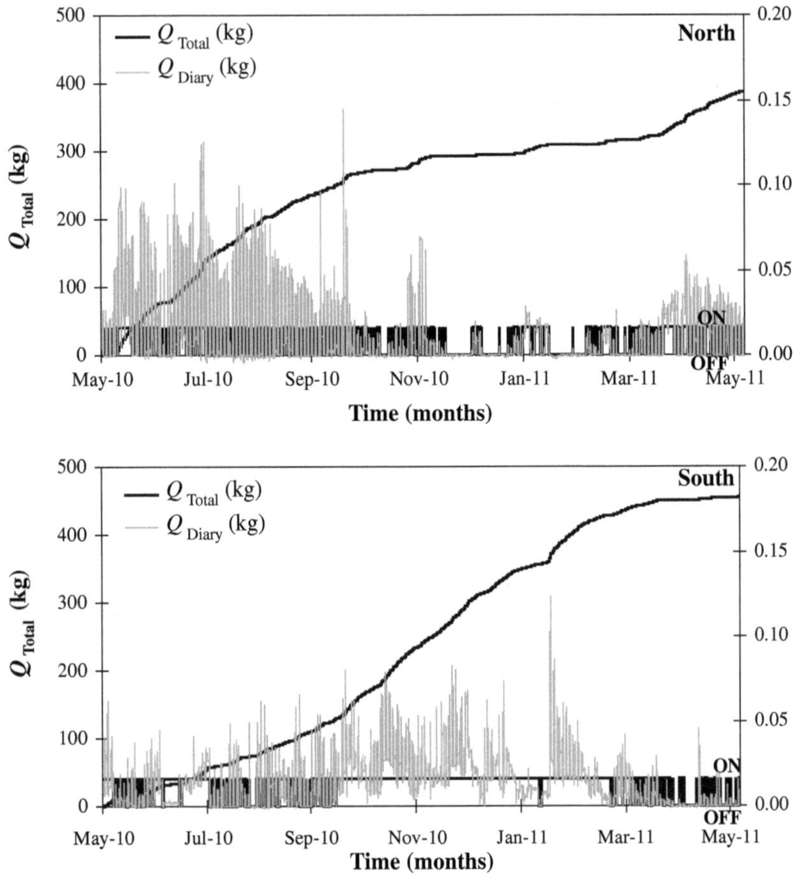

Fig. 12 Quantity of water vapour extracted of system. **a** North wall and **b** South wall

7 Conclusions

The main conclusions of our study, which set out to estimate the level achieved by the damp front, resulted on the following conclusions:

- The mechanisms of moisture transfer are complex, particularly in terms of rising damp in historical buildings. As rising damp is one of the main causes of decay in these buildings, it is important to study the factors related to this phenomenon.
- The simple analytical model for rising damp predicts values for the steady height of rise that are consistent with the numerical results, which provide a qualitative validation of the analytical model.

- Regarding the influence of material properties, the most important parameter is the absorption coefficient of water, which was as expected, since this parameter best characterizes the movement of water in liquid phase within construction materials.
- The results obtained, also, show that the steady-state height of rise increase with the wall thickness as a function of the square-root of the thickness.
- The evaporation potential, e, is an important parameter to be in consideration and the analytical results obtained were better in interior climates, as e is more stable.

In-field experiments in a historical church show that an efficient way of treating rising damp in historic buildings is to ventilate the wall base using the HUMIVENT device.

The analytical model of rising damp predicts steady height of rise figures that are consistent with in-field observations, which provide a good validation of the analysis. Using this model it is possible to scale the HUMIVENT system, optimise it and predict its behaviour. The numerical simulation results are similar to those obtained experimentally and analytically, which provide a qualitative validation of the results obtained.

References

Allen, R.G., Pereira, L.S., Raes, D., Smith, M.: Crop evapotranspiration (guidelines for computing crop water requirements). FAO Irrigation and drainage paper **56**, 290 (1998)

Allen, R.G., Smith, M., Perrier, A., Pereira, L.S.: An update for the definition of reference evapotranspiration. ICID Bulletin **43**(2), 1–34 (1994)

Bird, R.B., Stewart, W.E., Lightfoot, E.N.: Transport Phenomena, p. 187. John Wiley & Sons, NY (1960)

Bomberg, M: Moisture flow through porous building materials. Report no. 52, Division of Building Technology, Lund Institute of Technology, Sweden (1974)

Colombert, R.: L'Humidité des bâtiments anciens. Causes et effets; Diagnostic et remèdes, Paris, Editions du Moniteur (1975)

Crank, J: The mathematics of diffusion, 2nd edn. Oxford University Press, U.K (1975)

Evrard, A., De Herde, A.: Hygrothermal performance of Lime-Hemp wall assemblies. J. Build. Phys. **34**(1), 5–25 (2010)

Freitas, V.P., Guimarães, A.S.: Characterization of a hygro-regulated wall base ventilation system for treatment of rising damp in historical buildings. Proceedings of the 2nd Nordic Symposium on Building Physics, Copenhagen, Denmark, pp. 911–919

Green, W.H., Ampt, G.A.: Studies on soil physics, 1: The flow of air and water through soils. J. Agric. Sci. **4**(1), 1–24 (1911)

Guimarães, A.S., Delgado, J.M.P.Q., de Freitas, V.P.: Mathematical analysis of the evaporative process of a new technological treatment of rising damp in historic buildings. Build. Environ. **45**(11), 2414–2420 (2010)

Gummerson, R.J., Hall, C., Hoff, W.D.: Water movement in porous building materials-III. A sorptivity test procedure for chemical injection damp proofing. Build. Environ. **16**(3), 193–199 (1981)

Gummerson, R.J., Hall, C., Hoff, W.D.: Water movement in porous building materials-II. Hydraulic suction and sorptivity of brick and other masonry materials. Build. Environ. **15**(2), 101–108 (1980)

Hall, C., Hoff, W.D.: Rising damp: capillary rise dynamics in walls. Proceedings of the Royal Society A: Mathematical, Physical and Engineering Sciences **463**(2084), 1871–1884 (2007)

Hall, C., Hoff, W.D.: Water transport in brick, stone and concrete. Taylor & Francis, London (2002)

Hall, C.: The water sorptivity of mortars and concretes: a review. Mag. Concr. Res. **41**(147), 51–61 (1989)

Hall, C., Tse, T.K.M.: Water movement in porous building materials-VII. The sorptivity of mortars. Build. Environ. **21**(2), 113–118 (1986)

Hall, C., Hoff, W.D., Nixon, M.R.: Water movement in porous building materials-VI. Evaporation and drying in brick and block materials. Build. Environ. **19**(1), 13–20 (1984)

Hall, C.: Water movement in porous building materials-IV. The initial surface absorption and the sorptivity. Build Environ. **16**(3), 201–207 (1981)

Hall, C.: Water movement in porous building materials-I. Unsaturated flow theory and its applications. Build. Environ. **12**(2), 117–125 (1977)

Hanžič, L., Ilic, R.: Relationship between liquid sorptivity and capillarity in concrete. Cem. Concr. Res. **33**(9), 1385–1388 (2003)

Holm, A., Kunzel, H.M.: Two-dimensional transient heat and moisture simulations of rising damp with WUFI-2D, 2nd International Conference on Building Physics, pp. 363–3677. Belgium, Leuven (2003)

Kalamees, T., Vinha, J., Kurnitski, J.: Indoor humidity loads and moisture production in lightweight timber-frame detached houses. J. Build. Phys. **29**(3), 219–246 (2006)

Kay, A.L., Davies, H.N.: Calculating potential evaporation from climate model data: A source of uncertainty for hydrological climate change impacts. J. Hydrol. **358**(3–4), 221–239 (2008)

Kimball, B.A., Jackson, R.D., Reginato, R.J., Nakayama, F.S., Idso, S.B.: Comparison of field-measures and calculated soil-heat fluxes. Proceedings of the Soil Science Society of America vol. 40(1), pp. 18–25 (1976)

Kunzel, H.M.: Simultaneous heat and moisture transport in building components: One- and two-dimensional calculation using simple parameters. PhD Thesis, Fraunhofer IRB Verlag (1995)

Monteith, J.L.: Evaporation and environment, 19th Symposia of the Society for Experimental Biology, pp. 205–234. University Press, Cambridge (1965)

Oudin, L., Michel, H., Frederic, C., Perrin, C., Andreassian, V., Anctil, F., Loumagne, C.: Which potential evapotranspiration input for a lumped rainfall–runoff model? Part 2 – Towards a simple and efficient potential evapotranspiration model for rainfall–runoff modelling. J. Hydrol. **303**(1–4), 290–306 (2005)

Philip, J.R.: The theory of infiltration. Adv. Hydrosci. **5**, 215–297 (1969)

Schlichting, H.: Boundary-layer theory. Springer, Germany (2004)

Torres, M.I., de Freitas, V.P.: Treatment of rising damp in historical buildings: wall base ventilation. Build. Environ. **42**(1), 424–435 (2007)

Watt, D., Colston, B.: Investigating the effects of humidity and salt crystallization on medieval masonry. Build. Environ. **35**(8), 737–749 (2000)

Wilson, M.A., Hoff, W.D.: Water movement in porous building materials-XII. Absorption from a drilled hole with a hemispherical end. Build. Environ. **29**(4), 537–544 (1994)

Hygrothermal Performance and Damage Risk of Green Roofs

Daniel Zirkelbach, Beate Schafaczek and Hartwig Künzel

Abstract Green roofs become more and more popular especially in cities where they can help to improve the microclimate by reduced peak temperatures compared to conventional roof surfaces. They also reduce the gutter loads by rain water retention and save both heating energy in winter and cooling energy in summer. On the other hand, the cooler temperatures in summer may reduce the drying potential of the load bearing structure beneath the green roof. In inverted roofs the permanently humid conditions can lead to moisture accumulation in the foam insulation slabs. Due to a multitude of influencing factors like moisture storage, freezing, evaporation, shading and reduced radiation absorption through the plant cover etc. a simulation of the hygrothermal behaviour of green roofs represents a challenge. Therefore, the hygrothermal conditions of green roofs on wooden structures and on inverted concrete roofs at different locations has been measured and evaluated for several locations in Europe. Based on these experimental results a new approach was developed to calculate in a realistic way the hygrothermal performance of green roofs including the important influencing factors mentioned before. This contribution first investigates the conditions beneath the growth medium layers and the water content of the insulation boards and of the wooden sheathing by field tests in Holzkirchen, Vienna and Leipzig. In the second part a new approach to simulate green roofs including the growth medium layer is shown. In the case of inverted green roofs, the moisture film beneath the insulation boards is a dominant factor too that must be taken into

D. Zirkelbach (✉) · B. Schafaczek · H. Künzel
Hygrothermics, Fraunhofer Institute for Building Physics, Holzkirchen, Germany
e-mail: daniel.zirkelbach@ibp.fraunhofer.de

B. Schafaczek
e-mail: beate.schafaczek@ibp.fraunhofer.de

H. Künzel
e-mail: hartwig.kuenzel@ibp.fraunhofer.de

V. P. de Freitas and J. M. P. Q. Delgado (eds.), *Hygrothermal Behavior, Building Pathology and Durability*, Building Pathology and Rehabilitation 1,
DOI: 10.1007/978-3-642-31158-1_7,

account in the simulation. Finally, the results of simulations and experiments are compared in order to validate and fine-tune the new approaches that have been developed to model the behaviour of vegetated wooden structures and inverted concrete roofs in South, Central and North Europe.

Keywords Vegetated inverted roofs · Vegetated wooden flat roofs · Hygrothermal simulation · Thermal performance degradation · Wood decay

1 Introduction

Green roofs enjoy a good reputation especially in Europe. The aesthetic appearance of a green roof is normally perceived to be an advantage over other roofs. In summer time the high thermal inertia combined with evaporation effects helps to reduce the urban heat-island-effect. The plants improve the air quality and retention and storage of precipitation water reduce the gutter loads and thus compensate the urban soil sealing problems to a certain extent. Green roofs also provide some positive impacts on the building itself-due to their thermal inertia and the fusion heat of the retained water the temperatures beneath green roofs hardly fall below 0 °C in winter time. In summer evaporation cooling and self shading reduce overheating and cooling loads and thus improve the indoor comfort.

Green roofs are applied on different types of load bearing structures. Very common are concrete slabs with insulation and roofing membrane on top—which represents a more or less insensitive solution concerning moisture. Even slow moisture accumulations hardly impact the soundness of the construction as long as the roofing membrane remains impermeable to precipitation water. As the membrane normally is exposed to both extreme temperature and humidity variations and mechanical impacts during construction and operation, durability and water-proving may pose problems. Therefore a common alternative are so called inverted roofs. In such roofs the membrane is situated beneath the insulation slabs and thus is well protected from the mentioned climatic and mechanical impacts. The slabs are not glued or fastened to the substrate but ballasted with some cover layer like gravel, pavers or vegetated substrates. Drawback of this solution is the direct exposure of the insulation to moisture on top and bottom of the slabs. Only materials with a high vapour diffusion resistance shall be used to minimize the entry of moisture into the insulation. Inverted roofs are common and well proven assemblies which normally show good long term performance. Only in case of water retaining or diffusion retarding cover layers like green roofs or concrete slabs a slow accumulation of moisture can occur within the insulation panels. This can be explained by the prevailing boundary conditions: the rain water which penetrates the insulation at the board joints can lead to a permanent water film beneath the insulation slabs. Under these conditions the water content of the insulation increases in winter by vapour diffusion and condensation. In summer the condensate can only dry out when the relative humidity at the interface between insulation and cover layer decreases—this is the case for example for gravel layers while beneath a green roof growth medium or concrete pavers the relative humidity remains high most of the time also during summer. Considering the long-term moisture behaviour helps to specify the thickness of the insulation layer required to maintain the aspired level of thermal resistance over the whole service life of the roof.

Even more important is a reliable prediction of the detailed hygrothermal conditions when green roofs are applied to wooden structures. Normally it's a strict rule that wooden constructions need sufficient drying potential—ideally to both directions, inside and outside. Of course this is not possible in case of flat roofs, where the roofing membrane needs to be impermeable to liquid water and vapor diffusion. Thus drying is only possible towards the interior facilitated by solar vapour drive if dark colored membranes are used—at least in Central and Northern Europe with cold winter and moderately warm summer periods. Applying a green roof, results in a significant reduction of the peak temperatures during summer time which are essential for the drying by inverse diffusion towards the interior. Therefore wooden green roofs are less forgiving and show an increased risk of mould and rot decay at the exterior sheathing compared to normal flat roofs. Inadequate design combined with less-than-perfect workmanship can lead to moisture conditions which result in damage of the construction.

This means that green roofs on inverted insulation or wooden flat roof structures require a precise simulation of the hygrothermal conditions below the cover layers to allow a reliable evaluation of the thermal and hygric performance and the durability of these constructions. This is not possible with dew point calculations like Glaser. In the German moisture protection standard DIN 4108-3 the use of Glaser is explicitly discouraged for vegetated roofs—in such cases hygrothermal simulation represents the only alternative. Only these models include the important factors rain water absorption, liquid transport, moisture storage and latent heat effects for fusion and evaporation using hourly climate data with all relevant climate elements like solar radiation, wind and precipitation. However, up to now hardly any model for the use of green roofs in hygrothermal simulations was available in practice. This paper shows a new approach to simulate both, the sub structure as well as the single green roof layers including their moisture balance and the specific transfer conditions at the vegetated surface. For inverted roof insulations it is in addition necessary to represent the moisture conditions beneath the insulation slabs where water films can be observed caused by drainage of precipitation water. They are responsible for a big part of the vapour diffusion into the slabs, as the water film is on the warm side of the insulation and thus results in a high vapour partial pressure gradient beneath the insulation slabs.

Basis of the new model to simulate the green roof layers are field measurements on inverted roofs and on wooden structures at the field test site in Holzkirchen as well as on a test house of the MFPA Leipzig (Winter et al. 2007) and the Holfforschung Austria in Vienna (Teibinger and Nusser 2010). By the help of hygrothermal simulations with WUFI® Pro 5 (Künzel 1995) the conditions in the structures as well as in the green roof layers are simulated and compared to the field test results.

2 Field Tests: Measurements

2.1 Inverted Vegetated Roofs at the Test Field Site of IBP in Holzkirchen/South Germany (1985–2004)

The investigations on inverted roofs (Künzel and Kießl 1997) were carried out at the field test site of the Fraunhofer IBP in Holzkirchen, which is located on a plateau close to the Bavarian Alps in the south of Germany at 680 m above sea level.

On the concrete slab (sized 12 by 12 m^2) of a flat roof above a heated laboratory inverted roof assemblies with different composition have been installed (IBP-Report RB-17 1993) (Fig. 1). The individual test fields had a size of 3 by 3 or 6 m^2—allowing parallel investigation of eight different roof systems. The first tests started in 1985—the last ones were finished in 2004. Table 1 shows the different insulation thicknesses and cover layers above the sealed concrete ceiling for five of the roof systems. The longest investigation period lasted approximately 20 years.

Focus of the investigations was the hygrothermal behaviour of the insulation layer, i.e. the water content in the insulation boards and its influence on the thermal conductivity of the material. Therefore samples of the boards were periodically (in the main test period once or twice a year) taken from the roof and the water content of the boards was measured in the laboratory. To determine the hygrothermal conditions beneath the vegetated substrate—which is necessary to model the conditions in the cover layer later on—the temperature and humidity directly on top of the insulation were measured. For this purpose temperature (PT100) and dew point sensors (intermittently heated LiCl sensors) have been installed in the middle of the test fields as illustrated in Fig. 2 (IBP-Report HTB-2 1996).

The main test period lasted from autumn 1985 to summer 1992. Before disassembling the roofs in 2004 detailed moisture probings by weighing of the insulation boards was performed. The measured water contents are listed in Table 2 as average of normally two individual probing per year. The moisture content in all different test roofs showed a more or less continuous increase over the years. Depending on the different cover layers and material properties of the XPS boards the water content in the insulation boards varied between 4.8 and 8.7 Vol.-% (48–87 kg/m^3) after 19 years. The increase in water content is a function of the vapour diffusion resistance of the insulation material and the humidity conditions beneath and above the boards. In roof 2 the XPS Type B with lower vapour diffusion resistance reached a higher moisture content (8.7 Vol.-%) after 19 years than XPS Type A in roof 1 (4.8 Vol.-%). In roof 3, with EPS plant boards serving as water retaining and drainage layer, the water content of the XPS insulation reached 7.0 Vol.-% at the end of the test period.

As consequence of the rising moisture content, the thermal conductivity of the insulation material increases by approximately 2.8 % per Vol.-% (water content) according to (Zehendner 1979). This means that at the end of the test period the heat conductivity of the XPS boards reaches between 113 and 125 % of the dry state. This increase in conductivity should therefore be accounted for at the design

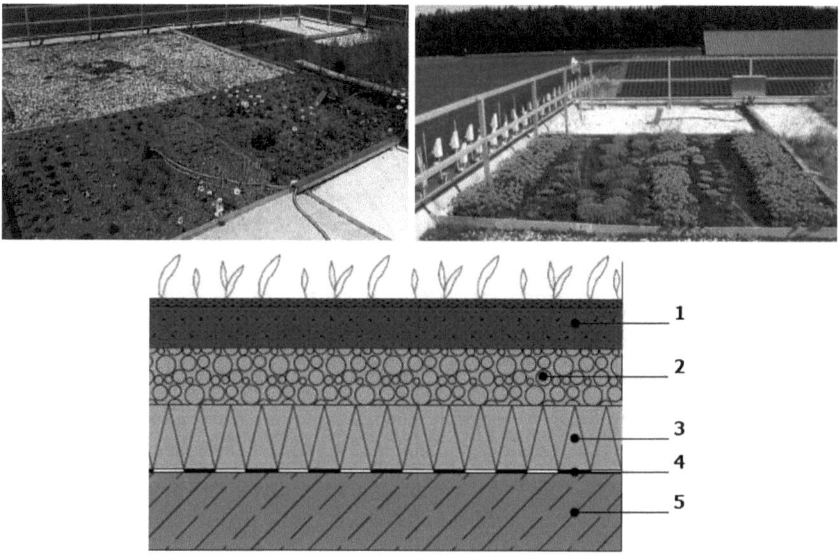

Fig. 1 Photographs of the green roof test fields with the dew point sensors seen from the East shortly after construction (*top left*) and a few years later from the South (*top right*). The building assembly (*bottom*) of roof 1 consists of 80 mm growth medium (*1*), 50 mm drainage layer (*2*), 100 mm XPS foam insulation (*3*), the roofing membrane (*4*) and 120 mm concrete ceiling (*5*)

Table 1 Test roof set up above concrete ceiling and roofing membrane

Roof	Construction
Roof 1 (1985–2004)	80 mm growth medium 50 mm drainage layer (gravel, expanded clay) filter fleece 100 mm XPS insulation Type A
Roof 2 (1985–2004)	80 mm growth medium 50 mm drainage layer (gravel, expanded clay) filter fleece 100 mm XPS insulation Type B
Roof 3 (1985–2004)	80 mm growth medium filter fleece drainage and water retention boards (EPS) 80 mm XPS insulation
Roof 4 (1995–1996)	80 mm growth medium filter fleece 50 mm drainage layer with gravel protection fleece 100 mm XPS insulation
Roof 5 (1995–1996)	80 mm growth medium filter fleece 50 mm drainage layer with expanded clay protection fleece 100 mm XPS insulation

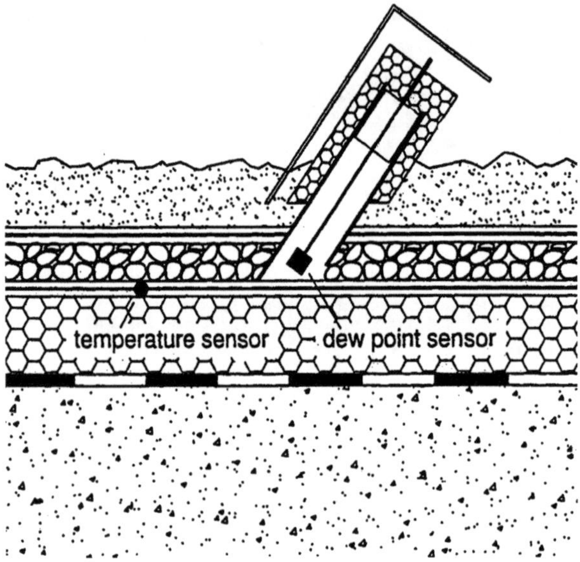

Fig. 2 Schematic figure of the dew point sensors in the drainage layer of the green roof

Table 2 Measured moisture contents in the different test roofs [Vol.-%]

Roof type	1986	1987	1988	1989	1990	1992	1996	2004
Roof 1	0.29	0.59	0.85	1.26	1.83	2.22	–	4.79
Roof 2	–	1.28	1.86	1.74	2.19	2.62	5.18	8.70
Roof 3	0.60	0.84	1.20	1.13	1.68	2.16	–	7.01

stage to allow correct U-value assumption and evaluation of the long term heat losses through such roof assemblies.

The measurements on roofs 4 and 5 were performed to get more detailed information about the temperature and humidity conditions beneath the growth medium of green roofs. The measured temperature beneath the gravel layer of roof 5 is plotted in Fig. 3. If compared to the ambient air temperature it shows a much more stable course in winter, i.e. the air temperature fluctuations are not followed. The values vary between minimum around −4 °C in winter and maximum around 35 °C in summer compared to −18 and 35 °C in the ambient air.

Concerning the humidity in the growth medium the results show, that the best drying conditions prevail in July and August when outdoor and roof temperatures reach their maximum values. The July of 1995 was particularly warm (monthly mean 19.2 °C) and dry (total amount of precipitation 40 mm) for the climate in Holzkirchen, while August (15.7 °C, 230 mm) came close to average conditions. Figure 4 shows the courses of the temperature (dashed line) and dew point temperature (solid line) at the exterior surface of the insulation in the different test fields and the daily amount of precipitation during these two months. Beneath the green roof the temperature and dew

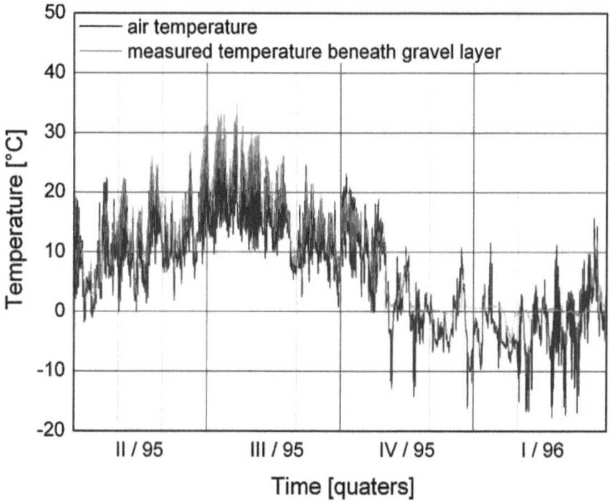

Fig. 3 Measured temperature beneath the gravel layer of roof 5 1995 and 1996

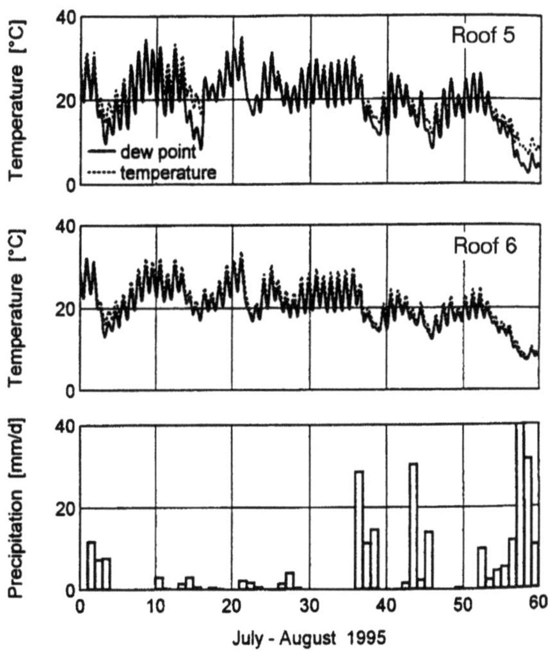

Fig. 4 Measured temperature and dew point temperature at the bottom of the drainage layers of the vegetated roofs 5 (gravel drainage, *top*) and 6 (expanded clay drainage, *middle*) together with the measured precipitation water (*bottom*)

point temperature coincide most of the time even during periods of very little rain. That means that the relative humidity there is close to 100 %.

Only if the dew point temperature falls below the measured temperature, the relative humidity falls accordingly below 100 % and signals drying conditions next to the insulation boards. The gravel in roof 5 does not retain the rain water to the same extent as the expanded clay and consequently dries out more quickly. This leads to some shorter periods where the dew point temperature falls a few degrees below the prevailing temperature which corresponds to RH values around

85 %. During these periods a certain dry out is possible which could explain the differences of the moisture contents in the boards of the roofs 1 and 2. Altogether the results show that beneath the growth medium the RH remains most of the time very high with values close to 100 %. Only without water retaining drainage layer the RH can fall to 85 % RH during very warm and dry periods in summer.

The measurements at IBP give information about the temperature and to a certain extent also on the humidity conditions under vegetated roof layers. Initial approaches to simulate green roofs based on these measurements showed that an agreement of simulated and measured values was only possible by adapting different material properties at the same time of the green roof layers. To create a reliable data set which would be transferable to different construction types and climate locations would therefore require further measurements and validations on different assemblies and at other climate locations.

2.2 Vegetated Roofs on Wooden Structures in Leipzig and Vienna (2008–2009)

In a common project between the Technical University of Munich and the material testing laboratory MFPA of Leipzig different wooden test roofs were built in Leipzig/Germany in 2007 including fields with vegetated cover layers (Fig. 5). The measurements comprise the temperatures beneath the roofing membrane as well as the wood moisture in the exterior OSB sheathing. The provided recordings for one of the investigated elements contained a period lasting from January 2008 to March 2009. The construction beneath 60 mm of mineral growth medium consists of a roofing membrane and a wooden structure with OSB sheathing on both sides (22 mm on top and 15 mm on the bottom side) and 220 mm cellulose fibre insulation between the joists. The moisture conditions in the growth medium as well as the surface temperature were not recorded, because the aim of the investigations was primarily to study the hygrothermal behaviour of the wooden structure in flat roofs with different insulation and vapour retarding layers. Detailed information on the investigations can be found in (Winter et al. 2007).

Similar focus had the investigation at the Austrian wood research institute "Holzforschung Austria" in Vienna. The different field test set-ups and results are published in (Teibinger and Nusser 2010). Here a mineral growth medium with 80 mm thickness was used on a wooden structure with roofing membrane on 20 mm OSB sheathing, 280 mm mineral wool insulation and a combination of a variable vapour retarder and a 16 mm vapour permeable wood fibre board as interior finish (Fig. 6). The measured data provided by Holzforschung Austria for one of the fields cover a time period from May 2008 to July 2009.

For both locations the indoor and outdoor climate conditions were measured with all required elements like wind, solar radiation and precipitation so that a comparative hygrothermal simulation becomes possible. The measurement results will be shown together with the simulation results in the following section.

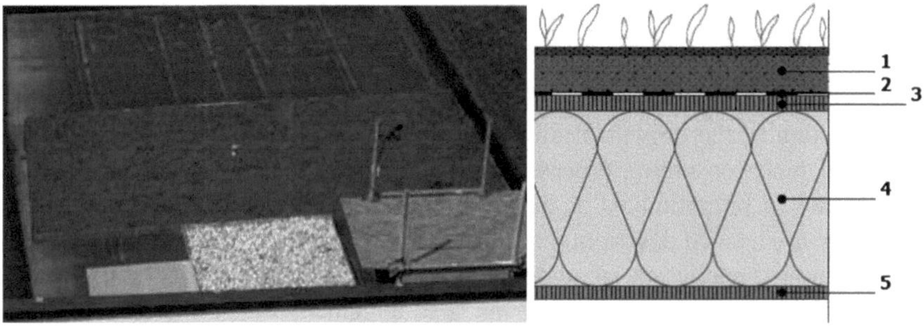

Fig. 5 Photograph (*left*) and roof assembly (*right*) of the test green roof field at the MFPA in Leipzig. The assembly consists of growth medium (*1*), roofing membrane (*2*), 22 mm OSB (*3*), 220 mm cellulose fibre insulation (*4*) and 16 mm OSB (*5*)

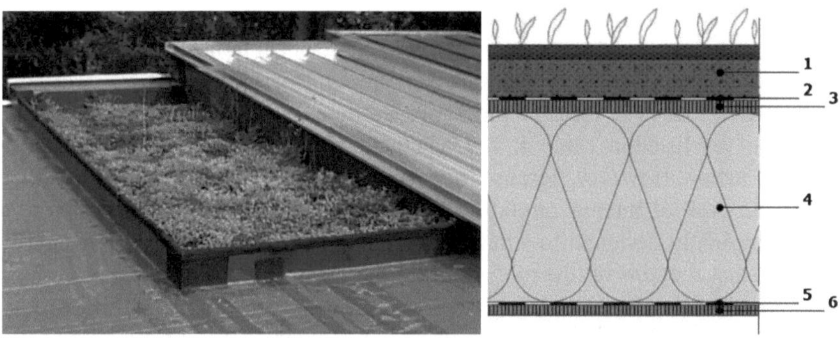

Fig. 6 Photograph (left) and roof assembly (right) of the test green roof field of the Holzforschung Austria in Vienna. The assembly consists of 80 mm growth medium (*1*), roofing membrane (*2*), 20 mm OSB (*3*), 280 mm mineral wool insulation (*4*), variable vapour retarder (*5*) and 12.5 mm gypsum board (*6*)

3 Hygrothermal Modelling of Green Roofs Including the Cover Layers

By the help of modern hygrothermal simulation tools it is possible to simulate the effects which are characteristic for vegetated inverted roofs and green roofs on wooden structures. While past calculations used constant humidity conditions above the insulation boards (Künzel and Kießl 1997), the following simulations include a new approach to model the hygrothermal performance of the vegetated growth medium and for the inverted roofs also the water films beneath the insulation slabs. The new cover layer model includes all relevant effects such as moisture absorption and storage, latent heat of fusion and evaporation and the reduced solar radiation absorption by partial shading provided by the plant cover. The simulations are performed with the hygrothermal simulation tool WUFI® (Wärme und Feuchte instationär—transient heat and moisture) which has been developed at Fraunhofer IBP for the past 15 years and validated by numerous laboratory and field tests (Künzel 1995).

3.1 Material Properties and Surface Transfer Settings for the Simulation

The adaption of the growth medium and the gravel layer properties was an iterative and time consuming process as many properties and surface transfer conditions had to be optimized at the same time and were influencing each other. Therefore no detailed description of the single optimization steps is given but only one resulting generic set of material properties, moisture sources and surface transfer coefficients which allowed the best agreement with all field tests. More detailed information can be found in (Zirkelbach et al. 2010, 2011).

The hygrothermal material properties of gravel were already available and determined by previous comparisons with field tests on inverted roofs with gravel cover layers; these are shown in Table 3. In the gravel layer no real capillary transport but only drainage effects can occur—therefore the rainwater in the simulation is not absorbed at the exterior surface but introduced over the whole thickness of the layer as a moisture source depositing 40 % of the particular precipitation load. The material properties for the growth medium of an extensive green roof are listed in Table 4. The vegetated substrate will absorb rain water by capillary action. However, because the precipitation water cannot be absorbed fast enough in case of intense rainfall, part of the rain water would be shed by the model originally designed to simulate the effect of driving rain; i.e. it would be assumed that it drains off the roof's surface the same way as driving rain drains of an exposed wall.

To avoid this effect which hardly occurs on a green roof in practice (the water will normally run off through the growth medium and not above it) again a moisture source is introduced which deposits further 40 % of the precipitation water into the bottom section of the growth medium.

Good agreement between measurement and simulation could be achieved by using a surface transfer coefficient of 19 W/m^2K and a very low solar radiation absorptivity of 0.3. The last value represents normally a bright surface which doesn't seem to agree with the optical appearance of the roof. That means that in the model the low short wave absorptivity considers not only the solar radiation itself but also some reduction due to self shading and long wave radiation losses. As the atmospheric counter-radiation was not measured for all three locations, the adaption of the setting was done without a detailed simulation of the long wave radiation exchange. Due to the high thermal inertia of green roofs the short term variations are of little relevance but the long term effects are considered by suitable values for surface transfer and radiation absorptivity—this mainly contributes to the low absorption coefficient of 0.3. With the different settings specified above a good agreement between simulation and measurement results was achieved likewise for all field tests. With slightly differing settings for the single roofs especially concerning the solar radiation absorption even better agreement would be possible—but the aim was a universally valid and safe solution for generic green roofs.

Table 3 Material properties for a gravel cover layer

Material property	Value
Bulk density	1400 [kg/m^3]
Heat capacity	1000 [J/kgK]
Heat conductivity	0.7 [W/mK]
Vapour diffusion resistance factor	1 [–]
Free water saturation w_f	50 [kg/m^3]

Table 4 Material properties for green roof growth media

Material property	Value
Bulk density	1500 [kg/m^3]
Heat capacity	1500 [J/kgK]
Heat conductivity	0.9 [W/mK]
Vapour diffusion resistance factor	5 [–]
Liquid transport coefficients DW0, DWS, DWW	1^{E-10}, 1^{E-6}, 1^{E-7} [m^2/s]
Free water saturation w_f	300 [kg/m^3]

3.2 Inverted Roof in Holzkirchen

The inverted roof in Holzkirchen was simulated using the material properties and settings described above. The material properties were taken from the WUFI® material database or derived from the information in the test reports (IBP-Report RB-14 1993; IBP-Report HTB-2 1996). Figure 7 shows the comparison of the measured and simulated temperatures beneath the cover layers of roof 5. The temperature curves are mostly running in parallel but a perfect agreement over the whole year could not be achieved—differences remain especially in winter when there is snow on the roof. A snow cover keeps the temperatures constant, close to 0 °C for longer periods while the ambient air temperature can vary around the freezing point. As the snow cover cannot be considered in the simulation, the resulting temperatures also show stronger variations around 0 °C. To keep the model on the safe side, the properties are adapted in a way, that the temperatures may fall below the minimum measured values in winter but should not exceed the maximum values in summer. The resulting difference can be observed for an extreme situation in February in Fig. 7 middle: the measured black curve remains around 0 °C while the simulated growth medium temperature falls to −10 °C and the air temperature even to −17 °C. In July the two curves fit quite well—the simulated curve is only slightly below the measured one. However the average agreement is very good. Over the whole year the average temperature difference only accounts for 0, 37 K—in winter it is -1, 13 and in summer -0, 84 K. In Holzkirchen the difference is somewhat larger compared to the other two locations. The reason lies in the thinner plant cover and a darker surface colour, which lead to higher surface temperatures. A better agreement would have been possible using a shortwave absorptivity of 0.4. However the long term conditions are better represented by a more dense plant cover as it could be observed in Leipzig and Vienna—therefore the lower value was also used in Holzkirchen.

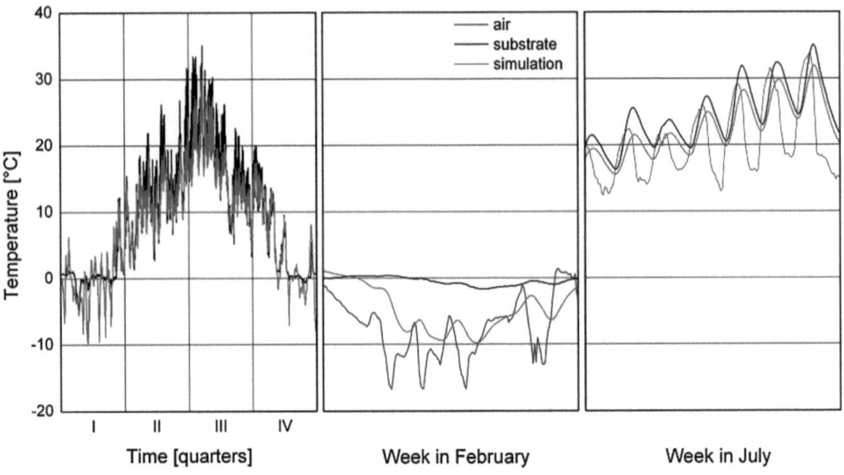

Fig. 7 Measured and simulated temperatures beneath the growth medium of roof 5 for the whole year (*left*) and for 1 week in winter (*middle*) and summer (*right*) in comparison with the exterior air temperature

To simulate the moisture content in the insulation slabs of the inverted roof two further points are of importance. First the water film between roofing membrane and insulation has to be considered in the simulation. Therefore an additional fleece layer with a further moisture source is modelled, that deposits 1 % of the precipitation water below the insulation. The source is limited to the free saturation of the fleece which corresponds to around 800 g/m^2. In Holzkirchen the RH in the fleece remains above 99 % RH all the year round with this source type. Also of importance is the inhomogeneous vapour diffusion resistance of the insulation boards. Due to the manufacturing process (extrusion of the polymer-gas mixture) the XPS boards have on both surfaces visible skins with significantly increased vapour diffusion resistance. In the laboratory a 10 mm thick surface layer of the boards was cut off to determine the different diffusion resistances of the core material and the skin (thinner layers are difficult to cut). In reality the extrusion skin is even thinner but with the assumption of a 1 mm thick extrusion skin with a rather high and a core with constant lower vapour resistance a good agreement between the measured and simulated data could be achieved.

Figure 8 compares the measured and simulated water content in the XPS boards of the roofs 1–3 at the field test site in Holzkirchen. The overall agreement is quite satisfying and the simulation as well as the field test results show a slow but continuous increase of the moisture content in the boards. The comparison of the model with further field investigation of inverted roofs at different locations in Germany, Switzerland and Austria confirmed the reliability of the approach and its transferability to locations with different precipitation loads.

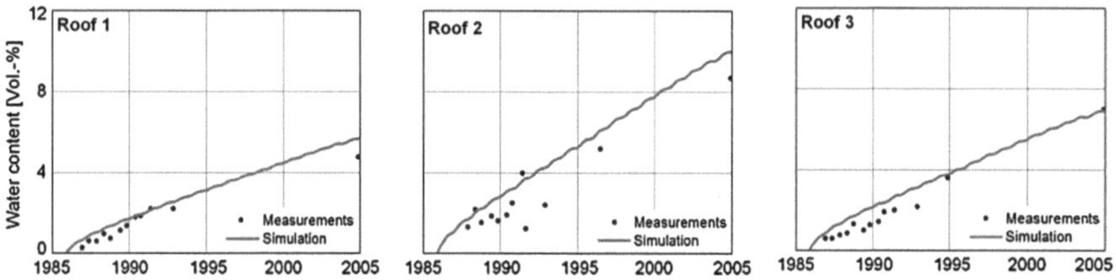

Fig. 8 Measured and simulated XPS water contents in the three test roofs at the IBP in Holzkirchen over the test period of approximately 20 years

3.3 Green Roofs in Leipzig and Vienna

The green roofs in Leipzig and Vienna are simulated with the same settings like the inverted roof in Holzkirchen. As climate data served the recorded meteorological data during the test period from both locations. As no inverted insulation is used, the fleece layers with moisture sources are not required. The material properties for the wooden structure were taken from the WUFI® material database—for the exterior OSB measured data were provided by the MFPA.

The comparison of the measured and simulated temperatures beneath the growth medium in Leipzig for the period from January 2008 to March 2009 is given in Fig. 9. Like in Holzkirchen a good agreement can be achieved during the most of the year and differences mainly arise when there is snow on the roof (January and February 2009). The difference of the average temperatures over several months accounts for only +0.26 K, −0.49 K in winter time and −0.15 K over the whole year. Thus the simulation comes very close to the measured hygrothermal conditions. Figure 10 shows the wood moisture of the exterior OSB in the roof in Leipzig. Here the water content in M % is compared until April 2011. Although the sensor recordings show some interruptions the simulated and measured results are very close to each other in the periods where all values are available. As well in the test as in the simulation a slow increase of moisture can be observed over the first years, which already indicates the problem. Most wooden green roofs experience some moisture increase during the first years after construction and too short test periods sometimes may not allow final evaluations of such structures.

In Vienna the agreement is even better. The two curves are plotted in Fig. 11. As only short periods with snow cover occurred in Vienna during winter 2008 and 2009 the differences between the curves remain very small. The coldest period at the beginning of January is still without snow—therefore also here the agreement between measurement and simulation is very good. The average temperature differences over the whole year amount to only -0.02 K (!); with −0.55 K in winter and +0.46 K in summer.

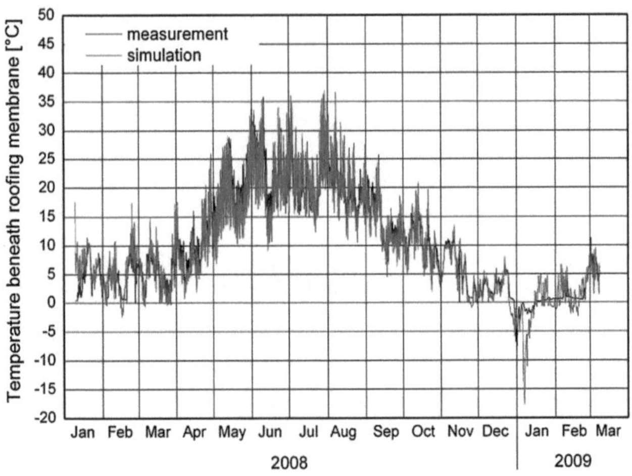

Fig. 9 Measured and simulated temperature beneath the growth medium at the wooden green roof in Leipzig for the test period from January 2008 to March 2009

Fig. 10 Measured and simulated wood moisture in the OSB board of the roof in Leipzig for the test period from January 2008 to April 2010

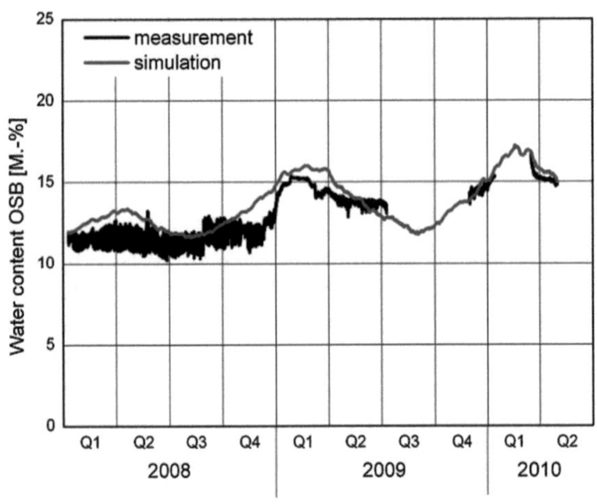

4 Long Term Behaviour of Green Roofs on Different Constructions

In summary it could be shown that the agreement between simulation and experiment for all three roofs at the three locations is very good considering all practical uncertainties. Therefore, the application of the generic green roof model for the analysis of different structures under various climate conditions seems justified. This will be done in the following section for inverted roofs concerning the risk of moisture accumulation in the XPS insulation boards and for wooden structures concerning the risk of rot in the exterior sheathing.

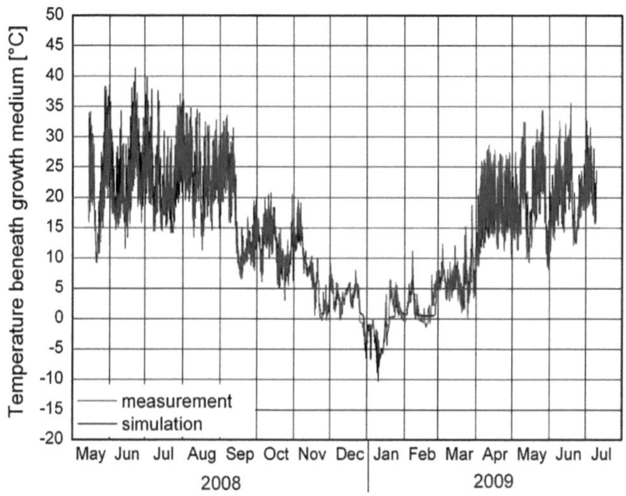

Fig. 11 Measured and simulated temperature beneath the growth medium at the wooden green roof in Vienna for the test period from Mai 2008 to July 2009

4.1 Inverted Roofs

The optimization of the simulation approaches showed the following main influencing factors for the moisture content in the XPS boards: the type of the cover layer, the diffusion resistance of the insulation material (skin and core), the outdoor climate and the temperature level below the roof. Therefore simulations with different outdoor climate, indoor temperature levels and cover layers are performed (Zirkelbach et al. 2011). As inverted roofs are also often used in combination with gravel layers, and the required material properties were available from previous investigations (not published), these results are additionally presented. As roof construction serves the same assembly as in the field tests in Holzkirchen. It consists of a 120 mm concrete slab which is sealed by a vapour impermeable roofing membrane and insulated with 100 mm XPS insulation boards above the membrane. The insulation layer is ballasted with either gravel or vegetated growth medium. As outdoor climate, weather files of the locations Espoo (close to Helsinki in Finland), Holzkirchen (Germany) and Lisbon (Portugal)are chosen to represent cold, moderate and warm European climate regions. For the three locations indoor climate conditions for residential buildings according to EN 15026 (2007) are derived from the local outdoor temperature. In Holzkirchen additionally a variation of the indoor temperature level (18, 22 and 26 °C indoor constant over the whole year) is performed to check the temperature influence on the water content in the boards.

Figure 12 shows the water content in the insulation boards of an inverted green roof at the three different locations. While the water content only increases moderately in Lisbon to about 2 Vol.-% after 30 years, the increase is much higher

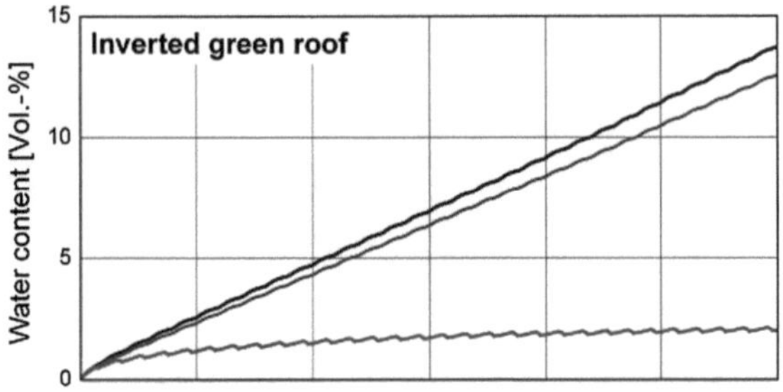

Fig. 12 Simulated water content of the insulation boards in inverted roofs with vegetation at Holzkirchen (Germany), Espoo (Helsinki, Finland) and Lisbon (Portugal) over a period of 30.

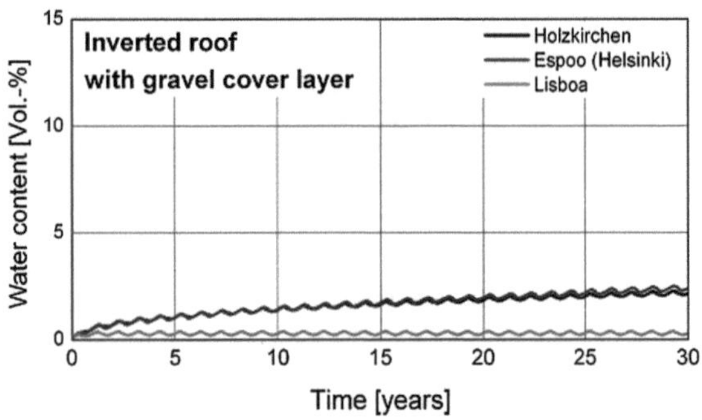

Fig. 13 Simulated water content of the insulation boards in inverted roofs with gravel cover layer at Holzkirchen (Germany), Espoo (Helsinki, Finland) and Lisbon (Portugal) over a period of 30 years

in colder regions with 12 Vol.-% in Holzkirchen and 13 Vol.-% in Espoo. In Holzkirchen and Espoo the diffusion flux from the water film beneath the insulation layer penetrates the boards from the bottom side and can hardly dry out to the top into the permanently humid growth medium. In Lisbon the temperature gradient over the insulation layer is smaller due to the milder winter conditions and the growth medium can temporarily dry out during the warm and dry summer period. Both effects together lead to a significantly smaller moisture uptake of the boards compared to the locations Holzkirchen and Espoo. The maximum water content in the boards beneath the gravel roof (Fig. 13) is less than 1 Vol.-% in Lisbon and around 2.5 Vol.-% in Holzkirchen and Espoo. For practical purposes this increase is insignificant and does not diminish the energy performance of the roof. The results for the gravel covered roof constructions are in line with practical experience and can be explained by the small water retention of the gravel layer. Thus it can also dry out in Espoo and Holzkirchen during periods with minor precipitation which facilitates the drying of the insulation boards.

The temperature level below the concrete slab is mainly responsible for the partial vapour pressure caused by the water film beneath the insulation boards and thus for the magnitude of diffusion flux into the XPS material. Figure 14 shows the water content in the insulation boards of green roofs in Holzkirchen for constant indoor temperatures of 18, 22 and 26 °C. While the 18 °C lead to approx. 10 Vol.-% after 30 years the values increase to 16 Vol.-% at 22 °C and even 25 Vol.-% at 26 °C. Again the conditions beneath the gravel covering are more favourable with only 1.5, 3 and 8 Vol.-% for the three temperature levels (Fig. 15). It should be

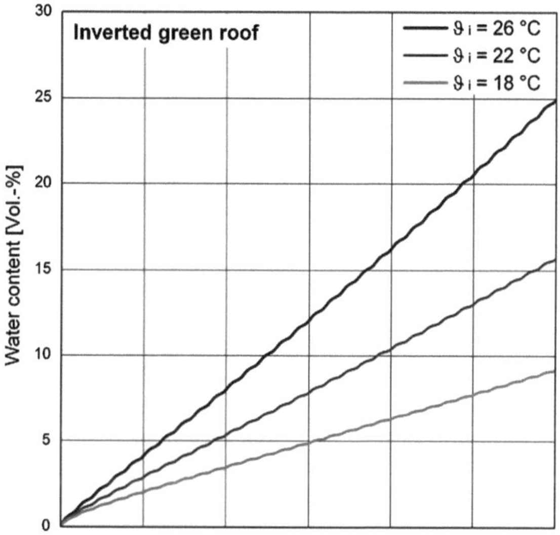

Fig. 14 Simulated water content of the insulation boards in an inverted roof with vegetation at Holzkirchen with different indoor temperature levels of constant 18, 22 and 26 °C

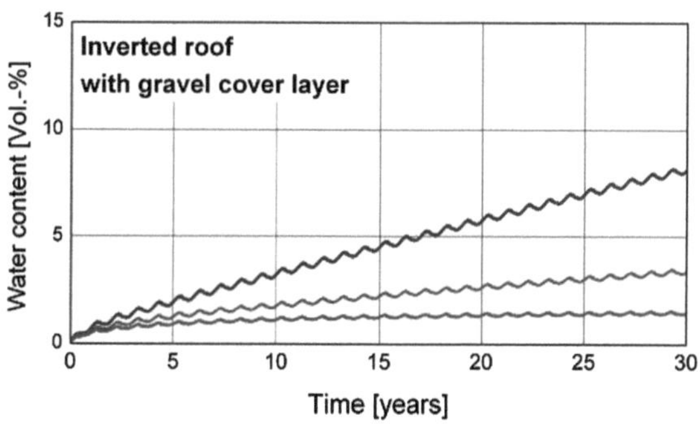

Fig. 15 Simulated water content of the insulation boards in an inverted roof with gravel cover layer at Holzkirchen with different indoor temperature levels of constant 18, 22 and 26 °C

noted, however, that indoor temperatures above normal room temperature may adversely affect the energy performance of all inverted roof assemblies even those with gravel cover.

4.2 Vegetated Wooden Roofs

The hygrothermal behaviour of wooden roofs depends on different factors—mainly the climate and exposure conditions, the type of the vapour retarder and the used insulation and sheathing materials (Zirkelbach and Schafaczek 2011). Measurement and simulation in Leipzig showed increasing moisture content in the exterior OSB-sheathing during the first 3 years. The maximum level of 17.5 M.-% reached during the third year is still uncritical concerning rot or wood decay. One

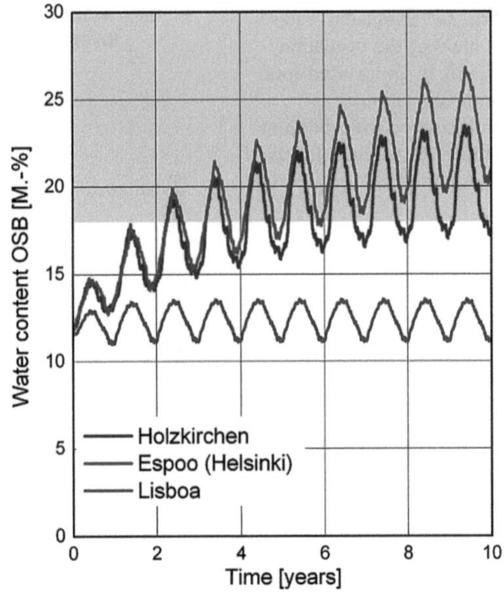

Fig. 16 Simulated wood moisture in the exterior OSB sheathing of the vegetated green roof at Holzkirchen, Espoo and Lisbon

explanation for the rising moisture could be the exceptionally cold winters during the test period. However the question arises how the wood moisture will continue under typical climatic conditions. Therefore the same roof structure was simulated with the reference year of Hannover which is located in a distance of approx. 200 km to the North East of Leipzig. These results—which are not plotted here show a further increase of the moisture content up to maximum values of 23 M.-% after 10 years. That indicates that also under "normal" climate conditions moisture levels similar to the ones observed in Leipzig have to be expected in the long run. This means the roof construction fails because the design limits of 18 resp. 20 M.-% for wood based materials or solid wood will be exceeded. Figure 16 shows the same structure exposed to the climate conditions of Holzkirchen, Espoo and Lisbon. As indoor climate normal conditions in residential buildings according to EN 15026 (2007) are used again. The results show a continuous increase of the wood moisture to values above 26 % by mass in Espoo. In Holzkirchen the water content increase seems to level off below 25 M.-% and only in Lisbon the drying potential during summer time is sufficiently high to keep the wood moisture below 15 M.-% all year round.

The results demonstrate that the wood moisture in the exterior sheathing of the investigated structure rises with colder conditions in winter and lower solar radiation loads in summer. This becomes particularly clear by comparing the conditions in Espoo and Holzkirchen. Both locations show similar annual average temperatures with 5.5 °C and 6.4 °C compared to 15.6 °C in Lisbon. However the remaining difference between the two locations is the significantly lower radiation load on a flat roof in Espoo with an annual sum of 870 kWh/a compared to 1223 kWh/a in Holzkirchen and 1713 kWh/a in Lisbon. A wooden structure which uses

Fig. 17 Simulated sheathing moisture content of vegetated green roofs in Holzkirchen. Variants include different exterior OSB sheathing materials and interior vapour retarding layers

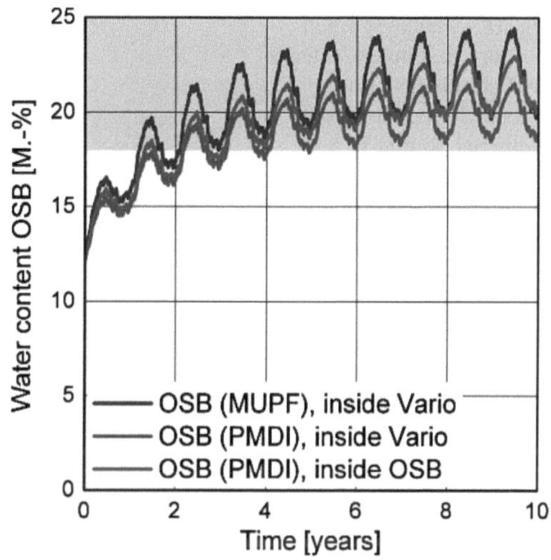

only a vapour retarding OSB at the interior side but no vapour retarder is therefore only suitable for regions with mild winters and warm summers—in moderate or colder climates they show an increasing risk of wood decay in the exterior wooden sheathing.

A second variation concerns the type of the OSB sheathing. The material properties of OSB vary significantly with the binder type, the production process and the density of the boards. In Central Europe a classification of OSB boards in different quality classes is common. In the classes 1 and 2 normally MUPF (Melamine-Urea-Phenol-Formaldehyde) binders are used which show a certain sorption capacity and low moisture resistivity. The more resistant binder type is PMDI (polymer diphenylmethandiisocyanate) which absorbs less moisture and is mostly used for the OSB types 3 and 4. While in the first construction in Leipzig no vapour retarder was used, the following variations have once again the OSB and for the other two cases a variable vapour retarder (polyamide film) with an s_d-value between 0.2 and 4 m on the interior side. The cases with variable vapour retarder include either PMDI or MUPF bonded OSB. The other construction details and boundary conditions are the same as before. The results for this variation are plotted as wood moisture content in the OSBs in Fig. 17. The OSB water contents in all variations exceed the critical design limits of 18 resp. 20 M.-%. With vapour retarder the maximum moisture content in the MUPF bonded OSB is approx. 3 M.-% higher than in the PMDI OSB. This shows that it can become important to use adequate material properties for the specific OSB which shall be used in practice. Different properties may result in different evaluations. With OSB instead of vapour retarder on the interior side, the water content in the PMDI OSB increases approximately by 1.5 M.-%. This is not surprising as the used OSB provides a lower variability of the vapour diffusion resistance than the variable vapour retarder—the Polyamide film is more diffusion resistant in winter to protect the structure and more permeable in summer time to allow a faster dry out.

Fig. 18 Simulated wood moisture in different exterior OSB sheathings of a vegetated green roof with additional 50 mm EPS insulation above the exterior sheathing (Holzkirchen climate)

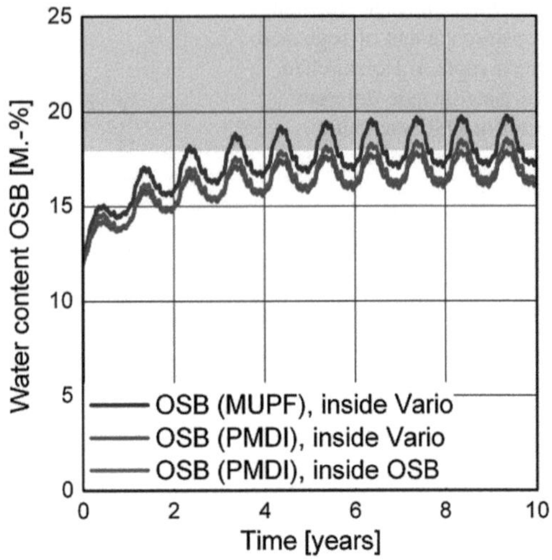

The initial simulations showed that green roofs with OSB or vapour retarder exceed critical design limits at colder locations like Holzkirchen and Espoo. This means wood decay in the exterior sheathing cannot safely be ruled out. The reason for the high moisture contents in the exterior OSB lies in the imbalance between vapour diffusion into the assembly in winter and dry-out in summer. Since the temperature moderating effect of vegetated growth media prevents better drying conditions, this imbalance may only be solved by reducing the amount of moisture getting into the roof in winter. This is possible by raising the temperature at the exterior sheathing, e.g. by applying additional insulation on the top of the roof—either in between wooden sheathing and roofing membrane or as inverted roof insulation beneath the ballast layers. In both cases the OSB remains warmer during winter time and thus experiences less condensation.

The same variants as shown in Fig. 17 are therefore investigated with additional 50 mm of EPS insulation above the OSB. The results are plotted in Fig. 18. The additional insulation reduces the water content in the different cases by nearly 5 M.-%. That means, that the MUPF OSB still exceeds 18 M.-% but remains below 20 M.-%. Depending on the specific national requirements this could be acceptable. The PMDI OSB with variable vapour retarder remains below 18 M.-% all year round, with OSB on the interior side the water content reaches only short peaks above 18 M.-% during winter time. For a few years now, a wooden flat roof guideline in Switzerland (Feuchteschutz bei Flachdächern 2007) recommends the additional insulation layer generally for green roofs to increase the moisture tolerance. The additional insulation should be applied over a vapour retarder installed on the exterior sheathing serving both as temporary weather barrier during construction and as protection against vapour diffusion into the upper insulation layer.

PMDI bonded boards can be generally recommended due to the high moisture level in green roofs during winter. The last variation concerning insulation type and thickness is therefore performed only with this OSB type—using the

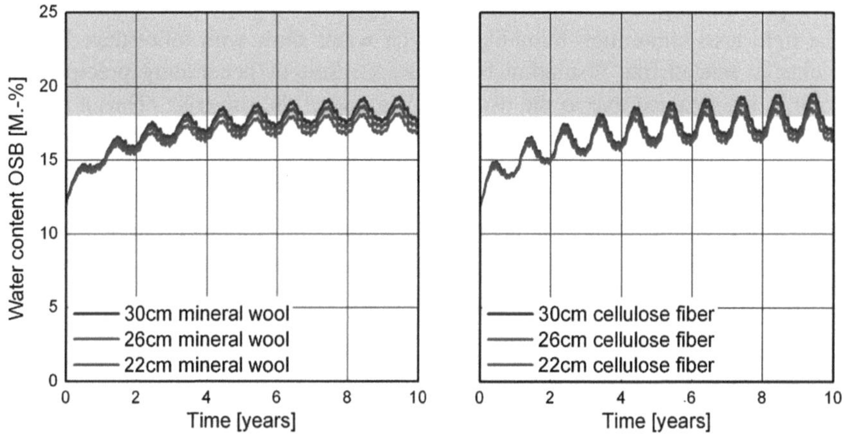

Fig. 19 Simulated wood moisture in the exterior PMDI-OSB sheathing of a vegetated green roof with additional 50 mm EPS insulation and different insulation types and thicknesses in Holzkirchen

construction from Fig. 19 with PMDI OSB on both sides. The insulation materials selected because of their difference in sorption behaviour are mineral wool (non hygroscopic) and cellulose fibres (with considerable moisture storage capacity). They are installed with thicknesses of 220, 260 and 300 mm which are common insulation levels in Central and Northern Europe. The resulting wood moisture in the exterior OSB shows slightly increasing moisture contents at higher insulation level. However, while the influence of the insulation thicknesses remains rather small (40 mm more insulation leads to approx. 0.7 M.-% higher moisture content). Only the OSB in the construction with 220 mm insulation remains below 18 M.-% in winter. This indicates that higher insulation levels on the lower side of the wooden sheathing also require a higher thickness of the insulation at the upper side. The type of insulation seems to be of little relevance—the long term behaviour of both mineral wool and cellulose fibre is very similar and they also reach the same maximum values during winter.

5 Summary and Conclusions

Green roofs offer many benefits concerning CO_2 control, particulate matter, microclimate, summer comfort conditions and aesthetics. However while their application on conventionally insulated concrete decks is rather straight forward some specifics have to be considered when vegetated growth media are installed on inverted roofs or wooden structures.

Inverted concrete roofs are well proven and reliable roof systems which provide high durability of the sealing layer as the insulation boards protect the roofing membrane from climatic and mechanical impacts. However the insulation boards are exposed to the outdoor weathering with high humidity and precipitation water. The field tests show high humidity or even water films with more than 10 mm thickness beneath the insulation boards as a result of penetrating precipitation water being retained due to the usual unevenness of the substrate. Thus a vapour diffusion flux into the boards cannot be avoided. Cover layers which can dry out during periods with little precipitation like gravel allow some drying of the boards towards the top of the roof which usually keeps the water content of the insulation low. Only in combination with high indoor temperatures in winter, for example in indoor swimming pools or buildings with industrial operation, the high vapour pressure beneath the boards results in a steeper increase of the insulation water content.

Cover layers which remain humid most of the time like those of green roofs, may provoke moisture accumulate in the insulation layer except in climates with warm and dry summer periods and mild winters. Under moderate and cold climate conditions especially in combination with higher temperatures beneath the roof, the increase of moisture cannot be neglected and the influence on the thermal conductivity has to be accounted for. Insulation materials with a higher diffusion resistance show smaller water contents than more permeable materials. The paper demonstrates that hygrothermal simulations can be applied successfully to predict the long term water content in the insulation and its influence on the thermal resistance of building assembly. Due to the wide range of measured data and many uncertainties concerning for example the unevenness of the underground care has been taken to assure that the simulation model—notably the moisture sources beneath the slabs—provides results which remain somewhat on the safe side. The simulation allows a detailed analysis of all influencing factors such as outdoor climate, indoor temperature and diffusion resistance of the insulation material. Thus the moisture increase may be reliably quantified which helps to maintain the aspired level of thermal resistance of the roof by adapting the insulation layer thicknesses.

The same hold for wooden structures where the simulations correspond well with measured temperatures and moisture contents in the exterior sheathing of the roofs. Notable differences only remain during periods with snow on the roof. However, these periods are of little relevance for the long term behaviour of the roofs and neglecting the influence of snow results in a slightly higher safety margin. A vegetated wooden flat roof represents a very sensitive solution which requires careful design and workmanship. An adaption to the specific outdoor climate is inevitable as the temperature level in winter and the solar radiation load in summer strongly influence the moisture behaviour. It is therefore obvious that shading of the roof (e.g. by trees, tall buildings or PV panels) can easily provoke damage and has to be avoided or accounted for. In all investigated roofs the wood moisture in the exterior OSB reaches levels only slightly below critical design limits. Apart from locations with mild winters and warm summers the application of an additional insulation layer above the exterior sheathing should therefore be considered. However at cold places and with high insulation levels beneath the wooden sheathing, the critical threshold may still be exceeded. Therefore general recommendations are difficult and a specific design analysis is required including notably the individual OSB and vapour retarder properties. The only solution for flat roofs that remains uncritical is to put the whole insulation layer on the top of the wooden structure—either as warm roof or as inverted roof insulation. In this case all wooden materials are located on the warm side and keep a humidity level close to the indoor climate.

References

EN 15026: Wärme- und feuchtetechnisches Verhalten von Bauteilen und Bauelementen—Bewertung der Feuchte-übertragung durch numerische Simulation, Deutsches Institut für Normung e. V., Deutsche Fassung (Juli, 2007)

Feuchteschutz bei Flachdächern in Holzbauweise: Merkblatt, Technische Kommission Flachdach, Gebäudehülle Schweiz (Herausgeber), Uzwil (2007)

IBP-Report HTB-2: Feuchverhältnisse in Umkehrdächern mit verschiedenen Deckschichten (Moisture conditions in inverted roofs with different cover layers), Holzkirchen, Germany (1996)

IBP-Report RB-17: Langzeituntersuchungen an begrünten Umkehrdächern (Longterm behavior of vegetated inverted roofs), Holzkirchen, Germany (1993)

Künzel, H.M., Kießl, K.: Moisture Behaviour of Protected Membrane Roofs with Greenery. In: CIB W40 Proceedings, Kyoto, Japan (1997)

Künzel, H.M.: Simultaneous Heat and Moisture Transport in Building Components. One- and Two-dimensional calculation using simple parameters. IRB Verlag, Stuttgart (1995)

Teibinger, M., Nusser, B.: Ergebnisse experimenteller Untersuchungen an flachgeneigten hölzernen Dachkonstruktionen. Herausgegeben von Holzforschung Austria. Wien (Forschungsbericht, HFA-Nr: P412) (2010)

Winter, S., Fülle, C., Werther, N.: Forschungsprojekt MFPA Leipzig und TU München "Flachdächer in Holzbauweise" (2007–2010)

Zehendner, H.: Einfluss von Feuchtigkeit auf die Wärmeleitfähigkeit von Schaumkunststoffen im Bereich von −30 bis +30 °C. Kunsstoffe im Bau **14**(H.1):18–22 (1979)

Zirkelbach, D., Schafaczek, B., Künzel, H.M.: Long-term hygrothermal performance of green roofs. In: American Society of Heating, Refrigerating and Air-Conditioning Engineers ASHRAE, Atlanta (Hrsg.), Thermal Performance of the Exterior Envelopes of Whole Buildings XI. Atlanta ASHRAE, 7 S (2010)

Zirkelbach, D., Schafaczek, B., Künzel, H.M.: Thermal performance degradation of foam insulation in inverted roofs to moisture accumulation. In: De Freitas, Vasco Peixoto (ed.) 12th International Conference on the Durability of Building Materials and Components DBMC. Conference Proceedings, vol. 1, pp. 529–536, Porto, FEUP (2011)

Zirkelbach, D., Schafaczek, B.: Gründächer im Holzbau Konstruktionen und Berechnungsverfahren. Tagungsband Bauphysik Forum 2011, Band 32 der HFA Schriftenreihe, S. 24–33 (April 2011)

Applicability of the Pull-Off Test: Teachings from a Large Sample of In Situ Tests

Nuno M. M. Ramos, M. L. Simões, João M. P. Q. Delgado and Vasco Peixoto de Freitas

Abstract The Pull-off test is a very popular way of evaluating the adhesion strength of renderings or ceramic tiles. Several standards frame the application of this test and interpretation of its results. In our laboratory, the in situ pull-off test is frequently applied to support the diagnosis of causes for building anomalies. The statistical meaning of the results, however, is clearly different from the one found in laboratory tests and hence reliability issues arise. In this work, 55 case studies based on in situ tests are analysed. The sample was retrieved from 15 test reports conducted by LFC-FEUP. In each of these reports different situations are included in terms of wall components, expected hygrothermal loads and aging history. This large number of tests supports a discussion on the applicability of the pull-off test regarding its reliability as a tool for two different subjects: decisions on repair strategies and building forensics. Typical variability found in field tests is described and analysed. The reliability of the test is analysed in detail with the help of one specific case where 40 individual pull-offs were conducted in the same building.

Keywords Pull-off test · Adhesion strength · In situ tests · Reliability

N. M. M. Ramos (✉) · M. L. Simões · J. M. P. Q. Delgado · V. P. de Freitas
LFC—Building Physics Laboratory, Civil Engineering Department Faculty of Engineering, University of Porto, 4200-465 Porto, Portugal
e-mail: nuno.ramos@fe.up.pt

M. L. Simões
e-mail: maires@fe.up.pt

J. M. P. Q. Delgado
e-mail: jdelgado@fe.up.pt

V. P. de Freitas
e-mail: vpfreita@fe.up.pt

V. P. de Freitas and J. M. P. Q. Delgado (eds.), *Hygrothermal Behavior, Building Pathology and Durability*, Building Pathology and Rehabilitation 1,
DOI: 10.1007/978-3-642-31158-1_8,

1 Introduction

The adhesive bonding of exterior revetments applied on façades is an important factor to ensure the safety and durability of the building. The failure of adhesive bonding has immediate consequences and therefore is a common concern for the building industry and building owners. The Pull-off test, classified as a near-to-surface, partially destructive method, is a classic way of evaluating the adhesion strength of renderings or ceramic revetments. Several standards frame the application of this test and its results interpretation. The pull-off test is frequently used as a tool to help on the correct diagnosis of causes for degradation of the façades exterior layers. The authors have applied its results in several cases and therefore use that hands-on knowledge as a starting point to address the subject of the test reliability.

Reliability of an engineering system can be defined with different terms. A common approach in structural engineering is to define reliability as the probability of successful performance, and so as the converse of the term probability of failure (Halder and Mahadevan 2000). But in the case of ceramic tiles detachment from a façade, for instance, the situation is more complicated. The behaviour of the system itself is highly uncertain due to several factors that make it heterogeneous by nature, ranging from substrate to workmanship. If a façade is under study, it is impossible to define its safety level, under the same standards that are usually applied in structural behaviour, using a semi-destructive technique such as the pull-off test. However, variability associated with the results of the in situ pull-off test should be addressed.

In this text, 55 case studies are analysed. The sample was retrieved from 15 reports of in situ inspections conducted by LFC-FEUP. This large number of tests supports a discussion on the reliability of the pull-off test for in situ inspections where variability of results is an important part of the decision process. A comparison with laboratory tests gives an idea of the different context where the results have to be interpreted.

The need for standards adapted to in situ conditions is evaluated. Existing standards are mainly applied in laboratory conditions, where variability will necessary be smaller.

2 Literature Review

2.1 Pull-off Test Variability

Ideally, when performing pull-off tests on a façade, the probability distribution of adhesive strength should be determined. This would allow for a supported risk assessment based on the estimation of the probability of failure. However, performing a large number of pull-offs would be economically unfeasible and could ultimately lead to the necessity of replacing the façade revetment.

The discussion on the acceptable variability is therefore of crucial importance. The variability of pull-off test results for in situ conditions was addressed by Bungey and Soutsos (2001) in a wider study on the reliability of partially destructive tests on concrete on site. The conclusions of that study pointed to the interest of using pull-off test on site since it's quick and straightforward to use and its results showed good correlation with cube compressive strength, but the lack of statistical confidence due to few in situ available tests was defined as a problem.

A study by Flores-Colen et al. (2009a, b) found, in a set of in situ tests, variation coefficients from 32 to 104 %. Several causes for the variability found included: using manual equipment with difficult control on load rate application, the aggressive conditions during test procedure and the difficulty to guarantee equal depth of the surface cores, equal glue thickness or nil load eccentricity in all cases. A correlation between causes and variability could not be precisely determined.

The study of pull-off test intrinsic aspects influence on variability of results was addressed by Costa and Carasek (2009). Those aspects included not only parameters already mentioned above but also geometry and dimension of samples. It was concluded that the tensile stress concentration that will develop in the borders of square samples by comparison with a more uniform distribution in circular samples results in higher adhesion values for the latter ones.

2.2 *Adhesion Failure*

A good knowledge of the causes for adhesion failure is important to define the most sensitive areas of a façade and hence reduce the candidate areas to test for searching the lower adhesion values. But the number of causes can actually be very wide as demonstrated by literature.

The causes of external wall tile failures have been widely reviewed by Ho et al. (2005). The author considered that the major causes are: the environmental effects (weathering, attacks and vibration and loading), the design and material factors (improper design and selection of materials specifications, joints and grouting and design and maintenance), the construction factors (improper surface preparation such as inadequate cleaning, no provision of proper keys, improper sequence of work, application/workmanship and adhesives/rendering) and the structural and substrate factors (substrate defects, substrate movement and substrate properties).

Other important common causes are: the differential movements between the tile, adhesive and the immediate substrates due to thermal, moisture or other effects; the failure of the cement rendering behind the adhesive; the structural movements, shrinkage and creep, vibrations and settlement problems and deformation of adhesive (or mortar) onto which the tiles have been laid due to shrinkage (Chew 1999; Mahaboonpachai et al. 2010).

Related to the environmental factors, some authors showed the influence of pollutant weather, temperature, relative humidity, wind, sunshine and air pressure, in the performance of tiles (Briffett 1991; Simpson and Horrobin 1970; Wallis

1995; Yiu et al. 2007). For example, Guan et al. (1997) demonstrated experimentally that high temperatures (an increase of 10 °C) accelerate the thermal degradation of polymeric materials in the adhesives. Briffett (1991) and Davies (1998) stressed the consequence of moisture penetration behind the tile surface through cracks. Experimental laboratory test carried out by Chew (1999) showed the influence of thermal induced and moisture-induced movements on adhesive strength. The author obtained an approximately 20 % reduction in pull-off strength after one 24 h thermal cycle in adhesives with polymer as additives, and a 6–37 % reduction in bond strength after wet and dry treatment.

The construction factors, such as workmanship, the tile setting pressure, the adhesive inner cavity and the adhesive open time were widely studied by Zhao and Zhang (1997). The authors showed that the pull-off strength decreases significantly when the setting pressure is below 50 % standard setting pressure; decreases proportionally with the increase of the adhesive inner cavity ratio; and decreases significantly when open time exceeds by 50 % the specified open time limit.

Related to the influence of the exposure temperature on the external wall tile failures, Chew (1999) demonstrated experimentally that the effect of temperature during application on the development of bond strength is significant. Application at 40 and 10 °C was found to have reduced the bond strength from 10 to 47 % compared to an identical application at 25 °C.

Finally, Maranhão et al. (2011) showed that moisture content above 6 % is sufficient to reduce 50 % of mortar deformability; the drying process allows the mortar to recover flexibility to a value similar to that prior to saturation; a logarithmic function gives the best fit for correlations between moisture content and flexibility and the water saturation increases mortar rigidity, as characterized by Young's Modulus.

2.3 Pull-off Test Standardization

The adhesive strength is determined as the maximum tensile strength applied by a direct load perpendicular to the surface being tested. The pull-off test is classified as a near-to-surface, partially destructive method which is able to measure tensile strength of various materials. The tensile load is applied by means of a defined pull-head plate glued to the test area. The adhesive strength is the ratio between the failure load and the test area. On all the tests described in this text, the test area was pre-cut in situ, fitting the size of the pull-head plates. The pre-cut was usually performed 5 mm deep inside the substrate or 5 mm beyond the interface under analysis. These could be square metallic plates of 50 × 50 mm with a thickness of 15 mm or circular metallic plates with diameter of 50 mm and 18 mm thick. In each case study, only one type of plate was used which means that results weren't affected by plate geometry variation inside each case study.

Fig. 1 Pull-off test machine

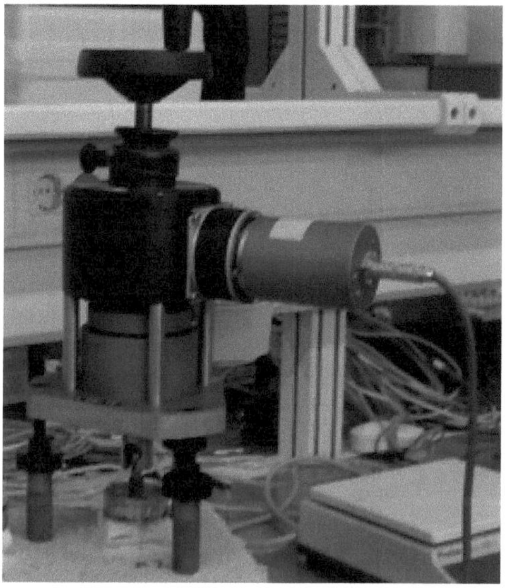

The test machine for direct pull tensile force test is in accordance with standard requirements, with the additional feature of automatic control of the applied force provided by an attached electrical engine, as presented in Fig. 1.

Several standards are available to frame the evaluation of adhesive strength by means of the pull-off test. The evaluation of tensile adhesion strength for cementitious adhesives, applied with ceramic tiles, can be found in EN 1348 (2007). The application of the pull-off test in cement based renderings is described in EN 1015-12 (2000). Both standards refer to laboratory tests and, therefore, don't focus on two subjects that are important for a successful implementation of the procedure in situ, namely pre-cut and minimum number of specimens.

The second aspect, crucial to the reliability of the test, is the amount of specimens needed and the admissible variability of the results. The EN 1015-12 (2000) defines a minimum of five specimens for the test. The EN 1348 (2007) also imposes a minimum of five specimens but the results must fall inside the range of ±20 % from the mean value. The RILEM. Recommedation MDT.D.3 (2004) defines six as the minimum number of test specimens to perform in situ tests on rendering but states that the final number should be higher if considerable heterogeneity is found.

The standard D 4541-02 (2002) is focused on pull-off test application to coatings. Although focused on paints and varnishes testing, an interesting approach is introduced as expected coefficients of variation are presented as reference for the application of the tests. The results were based in round-robin tests and pointed to intra-laboratory coefficients of 12 % and 20 % for inter-laboratory comparisons.

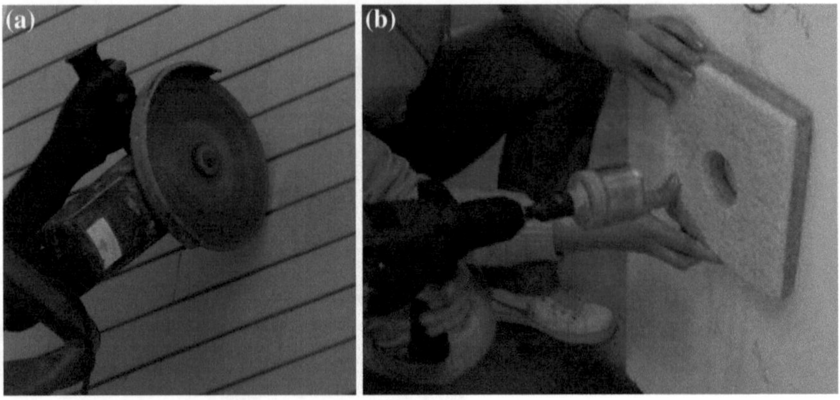

Fig. 2 Pre-cut of **a** ceramic tiles and **b** rendering

3 Pull-off Test In Situ Implications

3.1 Pre-cut Relevance

Performing a low disturbance pre-cut of each specimen is decisive for the implementation of the tests in situ. Usually, square specimens are preferred, as they are easier to produce. Another aspect is where to take the pre-cut. If it is performed deep in a render that's acting as substrate for the adhesive mortar, the failure will probably occur in the substrate, resulting in a false low result. It can however be interpreted as a performance indicator of the substrate and not of the adhesive. The pre-cut on the two different systems application of the pull-off test is presented in Fig. 2. In Fig. 2b a template is used to improve the quality of the pre-cut by reducing an eventual eccentricity of the core drill.

3.2 System Influence

The depth of the pre-cut is also of great importance. Depending on the system under analysis, the chosen depth will necessarily influence the interpretation of the pull-off result. Several examples are presented in the following figures.

In Fig. 3, a system of one layer rendering applied on masonry substrate is under test. In this case, the pre-cut reached the substrate, allowing for three different fracture patterns: adhesive fracture on the interface rendering/substrate, cohesive fracture on the rendering and cohesive fracture on the substrate. If one of the latter two occurs, the result is considered a lower bound limit for the actual adhesive strength.

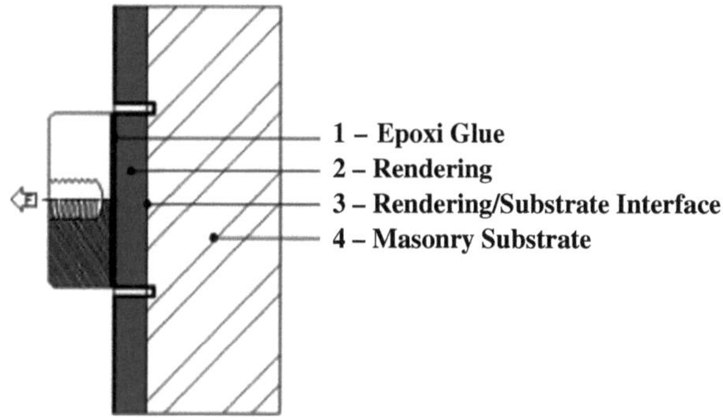

Fig. 3 Pre-cut of a one layer rendering applied on masonry substrate

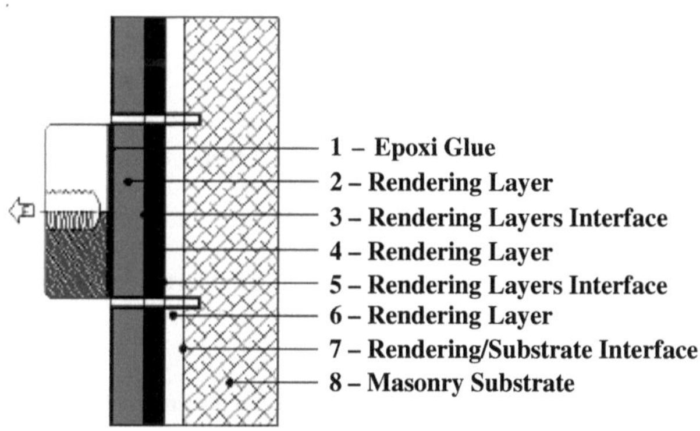

Fig. 4 Pre-cut of a three layer rendering applied on masonry substrate

The example presented in Fig. 4 show how the complexity can increase for the pull-off result analysis. In this case adhesive fracture can occur in the interfaces of the rendering's layers as well as cohesive fracture in each rendering layer and on the substrate.

Figure 5 presents a traditional rendering system, common in Portugal, where a lime rendering is applied over a bituminous impermeable layer. In this case, the difficult part will be deciding on the criteria for performance evaluation of such a system.

The typical configuration of ceramic tile application is presented in Fig. 6. Performing the pre-cut deep inside the substrate will again allow for different interpretations of the pull-off results. The fracture surface will indicate if the evaluation should be made on the performance of rendering or on the glue-mortar.

An example of pull-off application to ETICS (external thermal insulation composite system) is presented in Fig. 7. The pre-cut will usually be performed inside the thermal insulation, allowing for the interpretation of the system performance depending on the fracture mode.

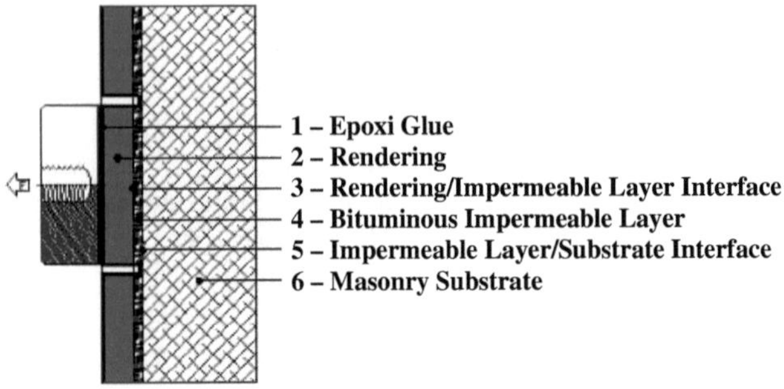

Fig. 5 Pre-cut of an old rendering applied over impermeable layer covering masonry substrate

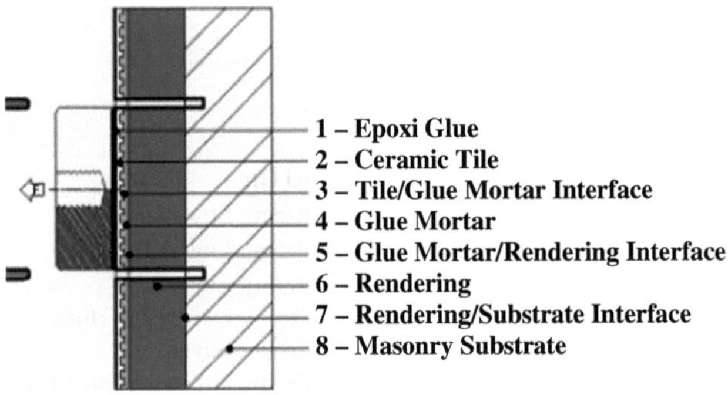

Fig. 6 Pre-cut of ceramic tiles applied with glue mortar over rendering applied on masonry substrate

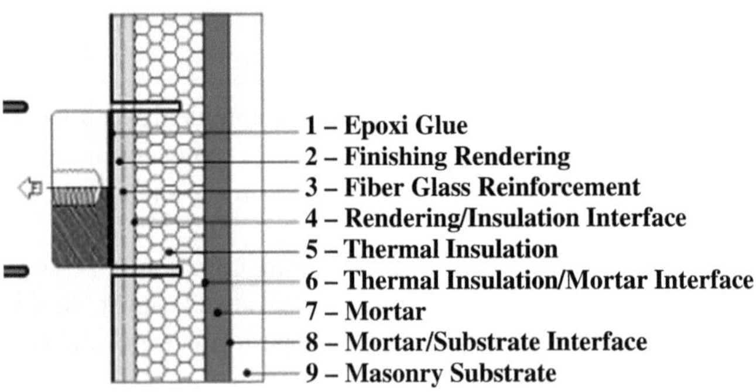

Fig. 7 ETICS pre-cut

3.3 Pre-cut Comparison

To assess the influence of the type of pre-cut a small exercise of comparison was made in three different wall finishing systems.

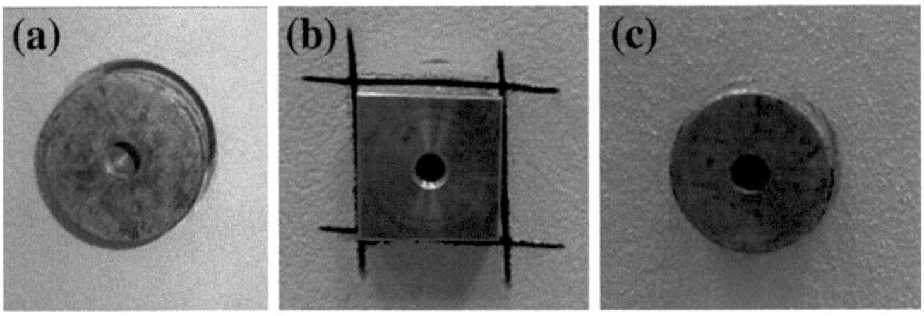

Fig. 8 **a** Circular pre-cut, **b** square pre-cut, **c** no pre-cut

Table 1 Pre-cut influence

Pull-off strength (MPa)	System A			System B			System C	
	Circular pre-cut	Square pre-cut	No pre-cut	Circular pre-cut	Square pre-cut	No pre-cut	Circular pre-cut	Square pre-cut
Mean	0.051	0.034	0.655	0.136	0.163	0.422	0.117	0.047
Standard deviation	0.019	0.019	0.079	0.035	0.011	0.063	0.022	0.019

The first system (system A) was a non conventional ETICS, where the insulation corresponds to cork plates and the finishing rendering is 1 cm thick including steel reinforcement. The second system (system B) was a one layer thermal rendering with EPS granules. The third system (system C) was a traditional ETICS.

Three types of tests were applied, varying the adopted pre-cut. The first type corresponded to a circular pre-cut, according to Fig. 8 a. On Fig. 8b, the square pre-cut is displayed and in Fig. 8c no pre-cut was adopted.

For each system, six pull-offs were performed, using the different pre-cut techniques. Table 1 presents the results of the tests in terms of mean and standard deviation. The results show that a difference between circular and square pre-cut is to be expected but the one will be higher will also depend on the type of system to be tested. The standard deviation values sowed no clear tendency to allow saying which is best. If no pre-cut is applied, much higher values should be expected. If the system is reinforced, which is true for System A, a value up to ten times higher can be found.

3.4 Access Conditions

To perform the pull-off test in actual in situ conditions is far more complicated than laboratory conditions which justify the increasing variability that can be found when both types of results are compared. Figure 9 illustrate the difficult conditions that were endured during an actual test.

The height where the test is performed is especially relevant as the increasing wind speed will introduce movement during the test and therefore increase the results variability.

Fig. 9 Difficult access to façades for performing pull-off tests

4 Experimental Study

This experimental study is based on data retrieved from inspections of buildings on operation phase. This implies that several factors conditioned the behaviour of the tested components such as the initial quality of the materials, their actual application conditions, substrate, façade exposure, age of the buildings and maintenance operations. The results of these tests cannot be directly compared since the conditions that lead to each set of results are unique.

The available data, however, presented an interesting opportunity to develop a statistical portrait of the adhesive strength that can actually be measured in revetments of building façades. The focus of the study is not only on the average values found but, more importantly, the spread of those values and the factors that influence it. The sample originated by the available data is random as it resulted

Table 2 Inspection reports and connected case studies

Inspection report	Case studies								Measurements	Material tested
	A	B	C	D	E	F	G	H		
1	5	4	4						13	Rendering
2	2	3	3	3					11	Rendering
3	5	3							8	Rendering
4	4	4	5						13	Rendering
5	2	5	2	7					16	Rendering
6	2	3	3						8	Rendering
7	3	3	3	3					12	Rendering
8	3	3	3	6	3	3			21	Rendering
9	5	5	4						14	Rendering
10	10	10							20	Cement based adhesive
11	3	3							6	Cement based adhesive
12	5	5							10	Cement based adhesive
13	4	5	5	5	5	5			29	Cement based adhesive
14	3	6	2						11	Cement based adhesive
15	5	5	5	5	5	5	5	5	40	Cement based adhesive
16[a]	4								4	Rendering
17[a]	3	3	3	3	3	2			17	Cement based adhesive

[a] The inspection reports refer to laboratory tests

from the study of buildings whose owners asked for inspections due to façade anomalies.

The results from 15 different in situ inspections were analysed. In each inspection, different zones in the building were tested, characterized by different components and/or different exposure. The data was therefore grouped by zones of identical conditions, allowing isolating more accurately the variables that contributed to the measured adhesion strength. The selected studies focused on ceramic tiles applied with cement based adhesives and cement based renderings as they were the ones with more data available.

Hence, the studied sample includes 232 measurements grouped in 55 case studies. All the invalid measurements were discarded from the statistical analysis. Table 2 presents the relation between the different tests and the original inspection report.

The grouping of case studies attended to the material that was actually tested. When testing in situ ceramic tiles adhesion the pre-cut is taken inside the mortar substrate. If the failure mode is cohesive inside the mortar instead of the adhesive the strength value must be associated with the mortar. Regarding substrates, concrete (7 % of the cases) and brick masonry (75 % of the cases) are typical solutions for buildings of the last decades while stone masonry (16 % of the cases) is common for older buildings. The only example of a non traditional solution included in this study was the application of ceramic tiles on wood-cement panels. Regarding substrate, there's a clear predominance in these tests of brick masonry, which corresponds to the Portuguese reality.

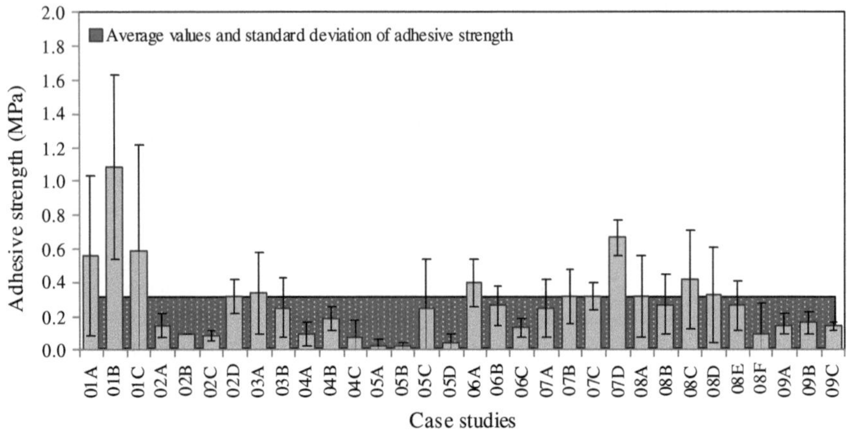

Fig. 10 Average values and standard deviation of adhesive strength for rendering case studies

The inspection reports 16 and 17 correspond to laboratory tests that were added to this study for reference comparison of variability. Report 16 refers to a straightforward test on rendering samples. Report 17, however, is a bit more complex test of a cement based adhesive. Case 17A is the reference test without aging while cases 17B to 17F are pull-offs after aging using different processes.

5 Results

An overview of the results retrieved from the 55 case studies is presented in Figs. 10,11, divided between rendering and cement based adhesives. Each case study is represented by the mean value and standard deviation. The graphs also include a reference value for each material, namely 0.3 MPa for renderings (NF P 15-201 1993) and 0.5 MPa for cement based adhesives (EN 12004 2001).

It can be observed that the mean values for adhesive strength of renderings were typically low and frequently below the recommended value of 0.3 MPa. The tested cement based adhesives exhibited higher strength and were frequently above the reference value of 0.5 MPa and even above 1.0 MPa. On both types of tested materials, significant variations of the standard deviations could be found. The low values observed, especially for renderings, could be related to the fact that these inspections were done in façades where anomalies could be observed.

6 Discussion

A deeper analysis of the adhesive strength found for renderings and cement based adhesives is presented in Figs. 12,13. The distributions found for each tested material were significantly different. While renderings presented a log-type

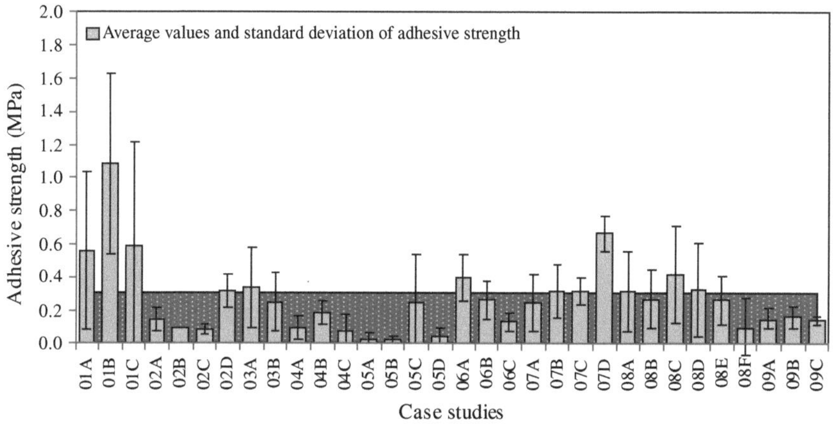

Fig. 11 Average values and standard deviation of adhesive strength for cement based adhesives case studie

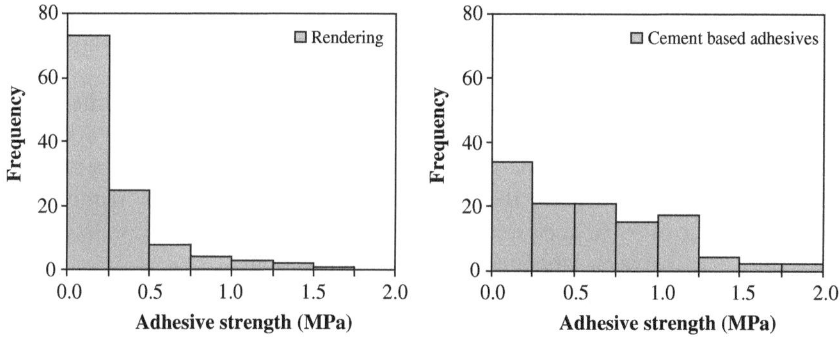

Fig. 12 Adhesion strength distribution for the tested materials

distribution with predominant results below 0.3 MPa, for cement based adhesives no theoretical model could be easily assigned. The mean was found above the reference value of 0.5 MPa. The box-plots for each material confirm this observation since, for cement based adhesives, no outliers were defined while for renderings, several values were defined as outliers, all corresponding to high resistance values.

The results from the laboratory experiments are presented in Fig. 14. They show not only higher average values but also, and more important, much lower standard deviations. This is important since the tests were performed by the same technicians and using the same equipment as the tests performed in situ.

A careful analysis of the data revealed that the laboratory test on rendering, case 16A, resulted in a coefficient of variation (CV) of 11.7 %. The highest CV in the laboratory tests on cement based adhesives, cases 17A to 17E, was 10.9 % in case 17A with a maximum difference to mean of 8.7 %, clearly below the 20 % limit

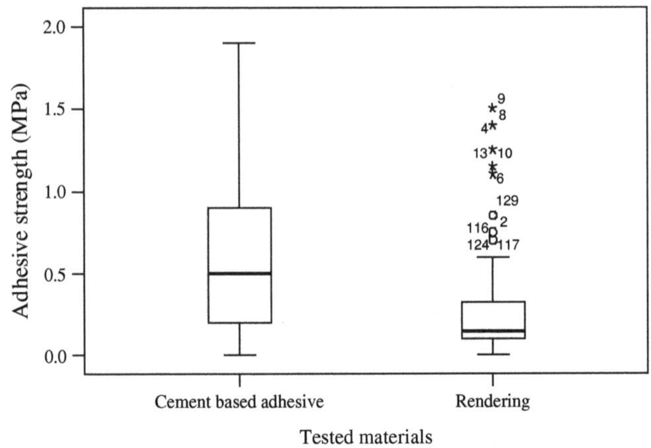

Fig. 13 Adhesive strength box-plots for the tested materials

defined in EN 1348 (2007). This means that the tests conducted in laboratory, presented as example, had low variability, within limits defined by standards, even for the test where aging processes were applied to samples.

Looking at the variability observed in situ, the conclusion is totally different. In the tests on rendering, cases from reports 1–9, the average CV was 67 % ranging to a maximum value of 173 % in case study 8F. In the tests on cement based adhesives, cases from reports 10 to 15, the average CV was 57 % ranging to a maximum value of 99 % in case study 15E. These values are quite revealing of the huge difference between laboratory and in situ conditions.

The analysis of the number of measurements influence on the variability of results is presented in Fig. 15, for the available results. The number of cases with 6 and 10 pull-offs was small, with only two for each, and so they aren't comparable with the results with the remaining cases. Looking at the results for cases with 2, 3, 4 and 5 valid pull-offs, the standard deviation of the coefficients of variation is smaller for 5 measurements. The correlation between the CV and the number of valid measurements seems to indicate that the greater the number of valid pull-offs the greater the probability of not missing the actual variability of adhesive strength under evaluation. The results indicate that the in situ CV for a specific set of pull-offs can range from 35 % to 80 %. The eventual lack of trust in a method that results in such high variability is controlled by standards as minimum values are imposed for single valid tests and not only to the mean value. The problem of in situ tests is that often the adequate number of valid tests is not achieved due to the obvious difficulties in repeating the tests.

The correlation between solar exposure and mean adhesive strength was tested for renderings. The results are presented in Fig. 16. The results weren't totally conclusive but still it could be observed that the values found on the East oriented façades were higher than the ones found on South and West oriented façades.

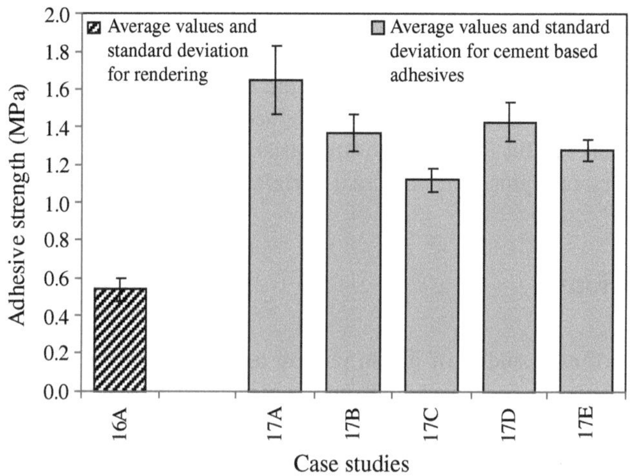

Fig. 14 Average values and standard deviation for laboratory tests

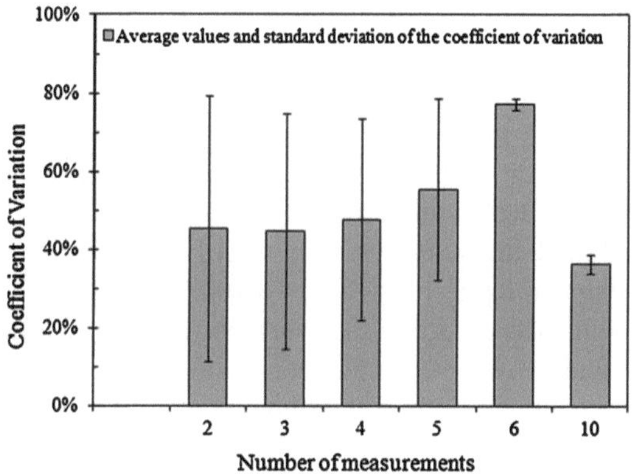

Fig. 15 Coefficient of variation correlation with the number of measurements per case study

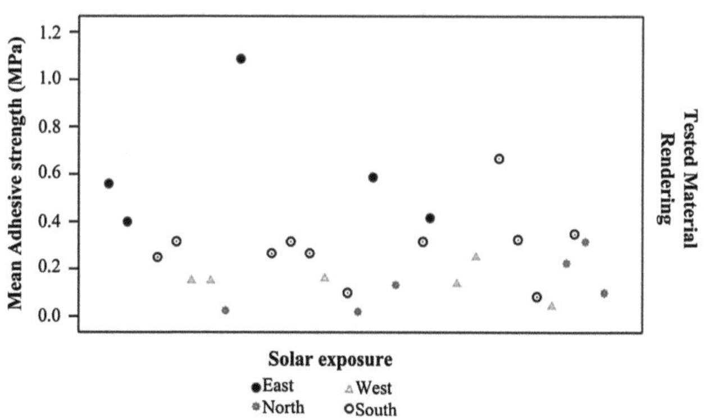

Fig. 16 Coefficient of variation correlation with the number of measurements per case study

Surprisingly, the North oriented façades had the lowest values. Except for this orientation, a correlation between surface temperatures in summer could almost be derived from these tests. The behaviour of the North façade may implie that other factors, such as wind driven rain or moisture accumulation would have to be taken into consideration. But for the exposure variable the sample is quite reduced for each cathegory implying that no actual correlations can be strongly supported.

7 Conclusions

The analysis of case studies of in situ pull-off test applications to adhesion strength evaluation retrieved from inspections to buildings resulted in the following conclusions:

- Pull-off test is used frequently as a decision support tool for building pathology studies;
- Adhesion strength can be measured for different systems and components according to international standards. These standards are however focused on laboratory tests. Documents for in situ measurements are scarce and not adapted to the entire range of components applied on façades;
- The adhesion strength distribution found for the overall tests on renderings was logarithmic, presenting many values below recommended resistance while for cement based adhesives no theoretical distribution could be assigned and the mean value corresponded to an acceptable resistance;
- The variability found in these tests was usually very high, corresponding to variation coefficients ranging from 40 % to 100 %;
- Regarding the influence of façade exposure on renderings adhesive strength, a tendency was found. However, the number of cases per orientation is too low to retrieve a definitive conclusion on this subject;
- A large number of pull-offs can lead to the definition of the adhesion strength distribution in a whole building with a good degree of confidence, leading to more robust decisions.

Although a large statistical sample was studied, the number of cases correlated to certain variables weren't sufficiently high to support definitive conclusions. Nevertheless, certain patterns were identified and the variability found for adhesive strength in situ tests inspires a cautious interpretation of results and a need for standards adapted to this specific subject.

References

AFNOR. NFP 15 201-1: Building works—Plasterworks made from cement mortars, lime mortars, mixed gypsum and slaked lime mortars (in French) (1993)

ASTM. D4541-02: Standard test method for pull-off strength of coatings using portable adhesion testers. ASTM International, West Conshohocken, PA (2002)

Briffett, C.: The performance of external wall systems in tropical climates. Energy Build. **16**(3–4), 917–924 (1991)
Bungey, J.H., Soutsos, M.N.: Reliability of partially-destructive tests to assess the strength of concrete on site. Constr. Build. Mater. **15**(2–3), 81–92 (2001)
CEN. EN 1015-12: Methods of test for mortar for masonry. Determination of adhesive strength of hardened rendering and plastering mortars on substrates, European Committee for Standardization, Brussels. (2000)
CEN. EN 12004: Adhesives for tiles—Definitions and specifications. European Committee for Standardization, Brussels. (2001)
CEN. EN 1348: Adhesives for tiles—determination of tensile adhesion strength for cementitious adhesives. European Committee for Standardization, Brussels. 2007
Chew, M.Y.L.: Adhesion of tiles for external cladding. Struct. Surv. **17**(1), 12–17 (1999)
Costa, E., Carasek, H.: Recommendations for the execution of the pull-off test in mortar renderings (in Portuguese), pp. 19–22. VII Simpósio Brasileiro de Tecnologia das Argamassas, Curitiba (2009)
Davies, H.: Repair methods for tile-clad buildings in Hong Kong. Struct. Surv. **16**(1), 34–38 (1998)
Flores-Colen, I., Brito, J., Branco, F.: In situ adherence evaluation of coating materials. Exp. Tech. **23**(3), 51–60 (2009a)
Flores-Colen, I., Brito, J., Freitas, V.P.: Expected render performance assessment based on impact resistance in situ determination. Constr. Build. Mater. **23**(9), 2997–3004 (2009b)
Guan, W.L., Alum, J., Liu, Z.J., Yang, T.: Performance of external tiled-wall systems under tropical weathering. J. Perform. Constr. Facil. **11**(1), 24–34 (1997)
Haldar, A., Mahadevan, S.: Probability, reliability and statistical methods in engineering design. Wiley, New York (2000)
Ho, D.C.W., Lo, S.M., Yiu, C.Y.: A study on the causes of external finishes defects in Hong Kong. Struct. Surv. **23**(5), 386–402 (2005)
Mahaboonpachai, T., Matsumoto, T., Inaba, Y.: Investigation of interfacial fracture toughness between concrete and adhesive mortar in an external wall tile structure. Int. J. Adhes. Adhes. **30**(1), 1–9 (2010)
Maranhão, F.L., Loh, K., John, V.M.: The influence of moisture on the deformability of cement–polymer adhesive mortar. Constr. Build. Mater. **25**(6), 2948–2954 (2011)
RILEM. Recommendation MDT D. 3: Determination in situ of the adhesive strength of rendering and plastering mortars to their substrate. Mater. Struct. **37**, 488–490 (2004)
Simpson, J.W., Horrobin, P.J.: The weathering and performance of building materials, 1st edn. Wiley-Interscience, New York (1970)
Wallis, K.: Buildings crumbling at rate of three each day. South China Morning Post, 20 Feb., Hong Kong, (1995)
Yiu, C.Y., Ho, D.C.W., Lo, S.M.: Weathering effects on external wall tiling systems. Constr. Build. Mater. **21**(3), 594–600 (2007)
Zhao, Z.Y., Zhang, W.L.: Influence of workmanship on the bonding strength of tiles to external walls. Int. J. Adhes. Adhes. **17**(1), 47–53 (1997)

Moisture Robustness During Retrofitting of Timber Frame Walls with Vacuum Insulation Panels: Experimental and Theoretical Studies

Bjørn Petter Jelle, Erland Sveipe, Erland Wegger, Sivert Uvsløkk, Steinar Grynning, Jan Vincent Thue, Berit Time and Arild Gustavsen

Abstract A large amount of the buildings in Norway is from the 1960s–1980s. Many of these buildings have timber frame walls and are now ready for retrofitting. Application of vacuum insulation panels (VIPs) may make it easier to improve the thermal insulation in timber frame walls with a minimal additional thickness. Retrofitting of timber frame walls using VIPs may therefore be performed without large changes to the building, e.g. extension of the roof protruding and fitting of windows. Additionally, U-values low enough to fulfil passive house standards or zero energy building requirements may be achieved, thus contributing to a reduction of the energy use and CO_2 emissions within the building sector. This work investigates different ways of retrofitting timber frame walls with VIPs on the exterior or the interior side. Timber frame walls retrofitted with VIPs on the exterior side is interesting because it allows for a continuous layer of VIPs over the building envelope, and it is also considered as a more robust solution than VIPs at the interior side (less risk of puncture). However, application of VIPs on the exterior side may cause condensation in the wall. To investigate this, a wall module containing four different wall fields was built between two climate rooms with indoor and outdoor climate, respectively. One field represents a reference wall built according to Norwegian building regulations from the 1970s. The three other fields represent different ways of improving the thermal insulation of the reference field, with VIPs at the interior or the

B. P. Jelle (✉) · E. Sveipe · E. Wegger · J. V. Thue
Department of Civil and Transport Engineering, Norwegian University of Science and Technology (NTNU), NO-7491 Trondheim, Norway
e-mail: bjorn.petter.jelle@sintef.no

B. P. Jelle · S. Uvsløkk · S. Grynning · B. Time
Department of Materials and Structures, SINTEF Building and Infrastructure, NO-7465 Trondheim, Norway

A. Gustavsen
Department of Architectural Design, History and Technology, Norwegian University of Science and Technology (NTNU), NO-7491 Trondheim, Norway

V. P. de Freitas and J. M. P. Q. Delgado (eds.), *Hygrothermal Behavior, Building Pathology and Durability*, Building Pathology and Rehabilitation 1,
DOI: 10.1007/978-3-642-31158-1_9,

exterior side. To minimize the size of the thermal bridge caused by traditional methods of fastening VIPs, a tailor-made VIP fastening bracket was applied in the build-up of the fields. Temperature, relative humidity (RH), and surface wetness was measured during the experiment. The surface wetness was measured on the wind barrier with a tailor-made surface wetness sensor consisting of double-sided tape, metal electrodes and paper sheets. In addition to the experimental investigations, numerical simulations and condensation control calculations were performed for the same wall fields with hygrothermal robustness performance as the main objective. In overall, the results from the experiments, simulations, and condensation controls conclude that timber frame buildings insulated with 100 mm mineral wool, might be retrofitted at the outside by adding 30 mm VIPs. However, this method for retrofitting provide limits to outdoor temperature, indoor moisture excess and indoor temperature.

Keywords Thermal insulation · Retrofitting · Timber frame wall · Vacuum insulation panel · VIP · Moisture · Robustness

1 Introduction

A large amount of the buildings in Norway is built in the 1960s–1980s, many of them in the 1970s (Statistics Norway 2009). The building tradition in Norway and Scandinavia implies that a large amount of these buildings have timber frame walls and are now ready to be retrofitted. The building regulations required in the 1970s a U-value of 0.35 $W/(m^2K)$ (about 10 cm mineral wool) in walls (Building regulations of 1969). Today the requirement is a U-value of 0.18 $W/(m^2K)$ (TEK 1997) and passive houses have even more strict requirements. That is, the allowed energy loss through a wall structure is close to halved in the past 40 years. Even if the new requirements do not have retroactive effect, the potential of saving energy is rather large in these old buildings. The CO_2 emissions from the building stock have lately been an issue for discussion. Investigations have shown that improving the thermal performance of buildings envelopes is one of the most economical ways to reduce CO_2 emissions (McKinsey 2009). This would also apply to the use of vacuum insulation panels (VIPs) since it leads to the same order of magnitude of environmental loading as conventional insulation materials (Binz et al. 2005). Kneifel (2010) has shown that thermal insulation together with low-emissivity windows, window overhangs, and daylighting controls can reduce the energy use by 20–30 % on average in new buildings in the U.S. It is therefore natural to assume that the reduction potential is even larger in retrofitting of buildings built according to older regulations.

VIP is a high performance thermal insulating material solution with thermal conductivity in the order of 0.004 W/(mK). The VIP is built up of a fumed silica core enclosed by a high barrier envelope to maintain the vacuum. More information on the

VIP build-up may be found in e.g. Willems et al. (2005), Brunner et al. (2006) and Tenpierik (2009). The ageing of VIPs is mainly caused by increased moisture and air content, i.e. a reduction of vacuum over time, where a thorough description of the ageing effects is found in Baetens et al. (2010a) and Wegger et al. (2011). Further work on thermal building insulation materials, including traditional, state-of-the-art and future insulation materials, may be found in Al-Homoud (2005), Papadopoulos (2005), Jelle et al. (2010) and Jelle (2011), and specifically more elaborated on gas-filled panels (Griffith et al. 1993 and Griffith 1995, Baetens et al. 2010c), aerogels (Schultz et al. 2005; Schultz and Jensen 2008; Hostler et al. 2008; Baetens et al. 2011) and phase change materials (Hasnain 1998; Farid et al. 2004; Demirbas 2006; Baetens et al. 2010b). A major challenge using VIPs is the risk of puncturing. A punctured panel has about five times higher thermal conductivity than an intact one, i.e. a conductivity of about 0.02 W/(mK) for a punctured VIP (Binz et al. 2005). In general, the use of VIPs requires a precise design of the structure and high accuracy in the craftsmanship because the VIPs cannot be adjusted or cut at the construction site.

Use of VIPs can make it easier to improve the thermal insulation in walls with a minimal additional thickness. Retrofitting of buildings using VIPs may therefore be done without large changes to the building, e.g. extension of the protruding roof and fitting of windows. At the same time, U-values low enough to fulfil the passive house standard may be achieved. In addition, many of the buildings from the 1970s have a need for changing the facades in any case.

Traditional advices for retrofitting timber frame walls are adding insulation on the outside of the existing wall. This normally prevents moisture problems because the temperature in the old wall structure rises. In addition, the thermal bridges in the old wall structure may be eliminated. However, this way of improving thermal insulation cannot uncritically be implemented using vapour tight VIPs. Introducing a vapour tight layer on the cold exterior side of a wall increases the risk of condensation inside the wall. However, VIPs have very good insulating properties and the temperature at the warm interior side of the VIPs might therefore be high enough to avoid condensation. SINTEF Building Research Design Guides (SINTEF 523.002, 2008) allow installing the vapour barrier 50 mm inside a 200 mm thick wall of regular mineral wool, e.g. ¾ of the total amount of insulation must be located on the cold side of the vapour barrier. The case of VIP applied on the cold side may be compared with such a case. Due to the differences in thermal conductivity, a thin layer of VIP on the outside of a mineral wool wall may still represent ¾ or more of the total thermal resistance of the wall. However, reduced thermal resistance of VIPs due to ageing or puncture may cause condensation in the structure and must therefore be accounted for.

This work investigates two different ways of retrofitting timber frame walls, one with VIPs on the cold side and one with VIPs on the warm side. A wall module containing four different fields was built. The wall was tested between two climate rooms with indoor and outdoor climate, respectively. The moisture and temperature conditions in the wall were logged and analysed. Furthermore, a tailor-made VIP fastening bracket and a proposed way of measuring moisture close to a

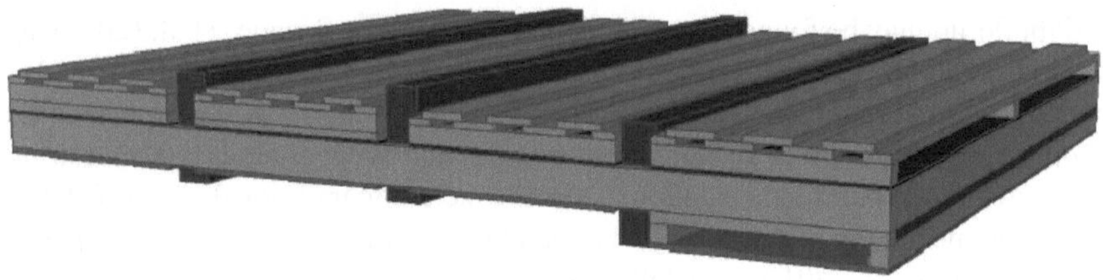

Fig. 1 The test module in 3D view, as seen from the bottom

material surface were applied. Calculations and simulations of thermal and moisture conditions in the different fields were also carried out and the experimental and theoretical results were compared and analysed. The work presented in this book chapter is an extended version of the work given by Sveipe et al. (2011a, b).

2 Experimental

2.1 Description of Test Fields

A full size realistic laboratory test may provide different results than a numerical analysis because of the complex mechanisms that a numerical model may not handle. The test module contains four fields as seen in Figs. 1, 2 and 3. Field 3 (F3) is built without VIPs and represents the original structure before retrofitting, i.e. a reference field of a timber frame wall from the 1970s. The three other fields (F1, F2 and F4) represent three different ways of improving the thermal insulation of the reference field (Fig. 2):

F1: Outside 30 mm VIP.
F2: Outside 20 mm VIP.
F3: Reference field.
F4: Inside 30 mm VIP.

The four fields are separated by plastic film and extruded polystyrene (XPS), i.e. separated with respect to moisture and heat transport. Figure 1 illustrates the test module with the four fields in 3D, whereas Fig. 2 shows a horizontal cross-section of the test module.

The build-up of Field 2 from outdoor to indoor is as follows (Figs. 2 and 3):

- Vertical wooden cladding.
- Furring strip (attached with a tailor-made fastening bracket, shown in the section 'VIP Fastening Bracket').
- VIPs, joint not taped (attached with the same tailor-made fastening bracket).
- Wind barrier (bitumen-impregnated paper, in common use in the 1970s).
- Mineral wool.
- Plasterboard.

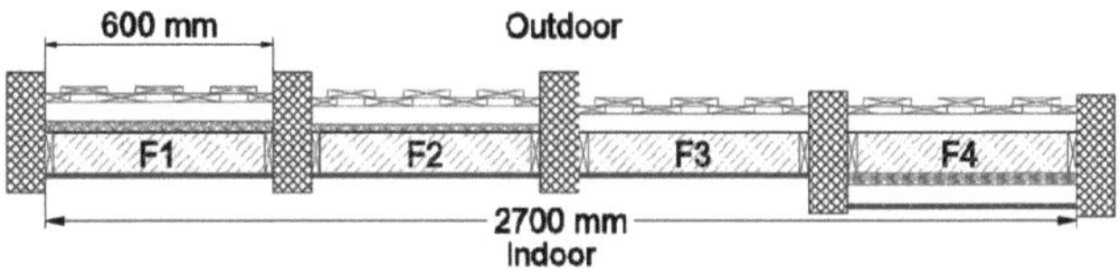

Fig. 2 Horizontal cross-section of the test module. Note that the fields have no vapour barrier

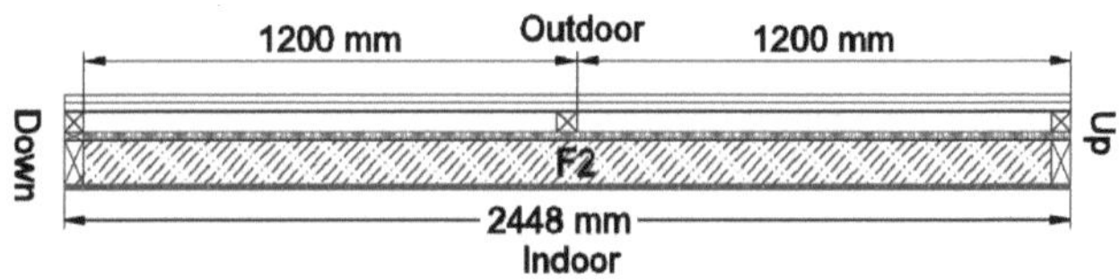

Fig. 3 Vertical cross-section of Field 2. Note that the field has no vapour barrier, only the vapour tight VIPs at the exterior side (F2 from Fig. 2)

Normally, the wall should have a vapour barrier as well, but this was omitted due to a relative short timeframe for the experiment. Not installing a vapour barrier represents a conservative modification and might be considered as a worst-case scenario.

The vacuum insulation panels used in the experiment in this work are of the type Vacupor NT-B2 from the producer Porextherm. Vacupor NT-B2 is a micro porous insulation material and consists of inorganic oxides. The main constituent is fumed silica. The other components are opacifiers for minimizing infrared radiation, and silicates. The panel is heat sealed with a high barrier film to maintain the vacuum (Porextherm 2009).

2.2 VIP Fastening Bracket

A high performance insulation material used the traditional way in timber frame walls (i.e. between wooden studs or furring strips) is not necessarily the best way to utilize the insulation materials' performance. Wooden studs between the VIPs will make thermal bridges and reduce the thermal performance of the wall. Development of new, or adjustment of old construction principles, may therefore be advantageous.

A tailor-made fastening bracket for VIPs was made in order to fasten the VIPs in an effective way with as low thermal transmittance as possible. The new VIP fastening bracket is made of a steel plate that is cut and bent so it can be fixed to the behind laying stud and at the same time hold the furring strip where the wooden cladding are fastened. This way the joint between the VIPs will be reduced to a minimum, i.e. the thickness of the steel plate. Figure 4 illustrates the dimensions of the steel plate before fabrication and model of the resulting

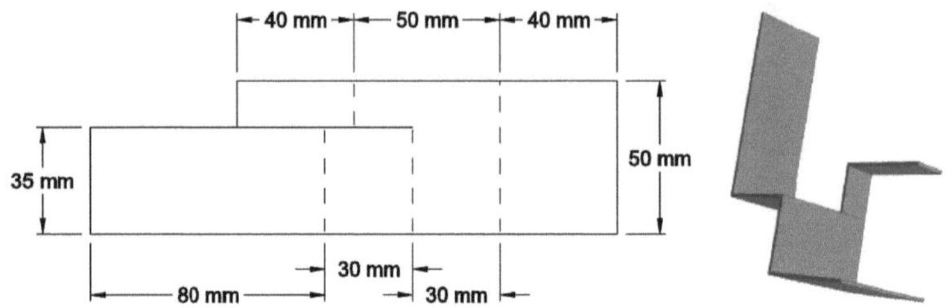

Fig. 4 One of the main VIP fastening brackets with dimensions where the *dashed lines* show where to bend (*left*) and a 3D model (*right*)

fastening bracket. The dashed lines in the drawing to the left in Fig. 4 show where to bend the brackets.

Figure 5 shows the application of the VIP fastening bracket in practise, depicting different steps during the implementation of the bracket. Note that the fastening bracket of steel was taped with a reinforced silver tape before installation in order to reduce the probability of VIP puncture due to sharp edges in the brackets. Other materials with lower thermal conductivities may be used for the fastening brackets, e.g. stainless steel, where one requirement is sufficient high mechanical strength, naturally.

A fastening bracket made of steel makes a thermal bridge in the structure. It was therefore important to address the size of the thermal transmittances and compare it to a traditional way of fastening the VIPs. Different versions of the fastening bracket, i.e. made of different materials, and a traditional way of fastening VIPs have been simulated in the 2D finite element program THERM (2003). The linear thermal transmittance (Ψ) of each thermal bridge version is calculated according to Eq. 1 (EN ISO 10211 2007).

$$\Psi = L_{2D} - \sum_{j=1}^{N_j} U_j l_j \qquad (1)$$

where

Ψ = linear thermal transmittance (W/(mK))

L_{2D} = thermal coupling coefficient obtained from a 2D calculation of the component separating the two environments being considered, including thermal bridges (W/(mK))

U_j = thermal transmittance of the 1D component, *j*, separating the two environments being considered (W/(m^{2}K))

l_j = length over which the value U_j applies (m)

The resulting average linear thermal transmittance of the horizontal joint between two VIPs is a result of the fastening bracket thermal transmittance (only present over the wooden stud) and thermal transmittance of the horizontal joint between two VIPs. Their influence is determined according to a ratio of their length and the total length. This means that for a timber frame wall with section

Fig. 5 Principle of the fastening bracket. The *top* photo shows the fastening brackets used in the middle of the wall (between two VIPs, fastened on a wooden stud), and fastening brackets used at the top plate. The three other photos show different stages in the installation of the fastening bracket between two VIPs

width of 600 mm (from stud to stud) and 36 mm studs, the part of the average thermal transmittance that relates to the fastening bracket is (0.036 m)/(0.6 m) and the part of the average thermal transmittance that relates to the joint between two VIPs is (0.564 m)/(0.6 m). Thermal conductivity of the materials used in the simulations is mainly collected from EN ISO 10456 (2007). Material data for the fastening brackets enveloped with aluminium tape was assumed to be 25 and 9 W/(mK) for tape-enveloped steel and tape-enveloped stainless steel, respectively, by area weighing calculations. The results from selected thermal transmittance simulations are described in the Results section 'VIP Fastening Bracket'. The U-values of Field 1 and Field 3 have been calculated according to EN ISO 6946 (2007) applying the thermal transmittance found.

2.3 Surface Wetness Sensor

In order to be able to measure condensation on a surface a tailor-made moisture sensor was developed (Fig. 6). An air humidity sensor may show a relative humidity (RH) below 100 %, while condensation has actually occurred on the

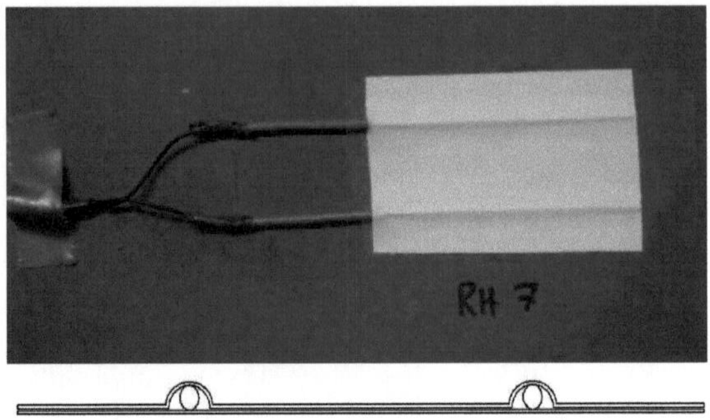

Fig. 6 The wetness sensor consists of double-sided tape, single-wired cable and a paper sheet (cross-section drawing at bottom)

adjacent cold surface. The principle is to measure the electrical resistance on a thin material taped on the respective surface, i.e. the measurement will be as close to the material surface as possible. Different materials for use in the moisture sensor were considered and tested. Finally, the choice was a regular copy paper (Lyreco Budget, 80 g/m^2). The sensor consists of double-sided tape, plastic-insulated single-wired 1.5 mm^2 cables, and a sheet of copy paper. The moisture sensor is henceforth called the *wetness sensor*, and is depicted in Fig. 6.

The electrical resistance of the wetness sensor was measured with a wood moisture content meter, i.e. the electrical resistance was expressed by the moisture content in spruce (at 20 °C). The sensor was calibrated in advance by measuring the moisture content values for different RH in a climate chamber at 10 °C. The result of this calibration made it possible to convert the measured moisture content from the experiment to RH by a linear function on the form $y = ax + b$, where the 'a' and 'b' values were calculated. The moisture content measured for the case of condensation on the sensor was also noted. However, this method provides for several unknown errors. The calibration curve that describes the *electrical resistance's* connection to *moisture content* for paper is not necessarily in linear equivalence with the calibration curve for spruce, even if this has proven to correspond quite well. The fact that the electrical resistance and hence the calibration curves change with temperature is not accounted for (the calibration was done at 10 °C). It is known that wood moisture meters are most accurate at 7–25 % wood moisture content by weight (Geving et al. 2006). The wood moisture content measured at a wet wetness sensor was about 25–30 %, which means that the wetness sensor measurements are less accurate from about 100 % RH. In this range, a small change in the electrical resistance causes a large change in moisture content. This may also be seen in the later depicted results from the wetness sensor. In addition, the size of errors of the manufacturing and the errors of materials used in the wetness sensor is not known. Therefore, the wetness sensor is to be considered as an indicator of RH and condensation. The sensor has to be further verified before it is considered to measure RH values. Especially the calibration work should be investigated further.

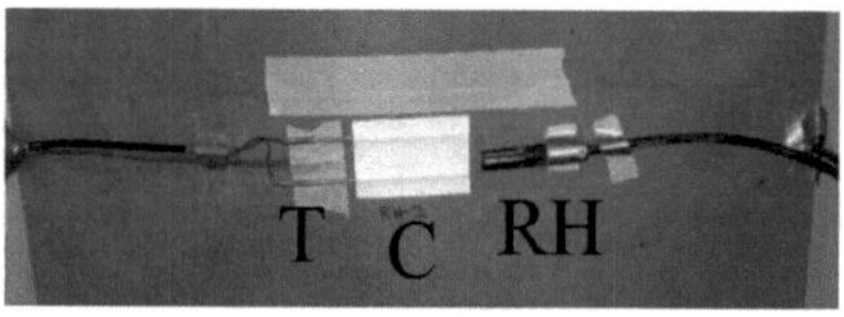

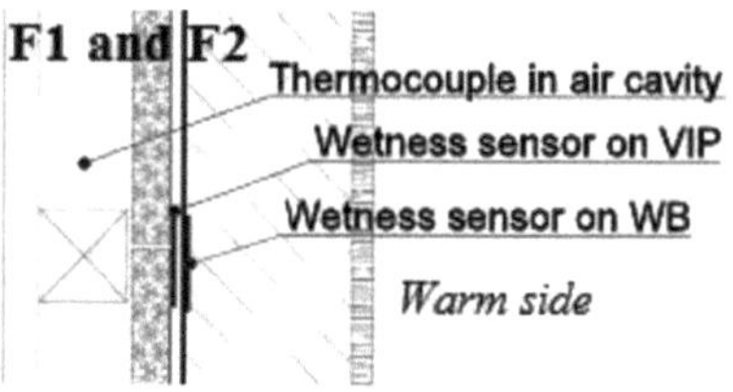

Fig. 7 *Left* photo depicts the thermocouple (T), surface wetness sensor (C), and relative air humidity sensor (RH) as installed on the wind barrier before the mineral wool was added. Right drawing shows a cross-section of F1 and F2 with a wetness sensor both on the wind barrier and on the joint between two VIPs at the middle of the fields. The thermocouple in the air cavity is also shown, while other sensors on the wind barrier (WB) are not depicted

2.4 Temperature and Moisture Measurements

Altogether, 36 sensors of three different types were used in experiment (Fig. 7):

- Thermocouples (T)
- Relative air humidity sensors including thermocouples (RH)
- Surface wetness sensors (C)

The three different sensors are depicted in Fig. 7. The thermocouples consist of a cable with two separate conductors (copper and constantan). The cables are twined and soldered together at the tip where the measurements based on varying electrical potentials take place. The air humidity sensors were of the type Vaisala humidity and temperature transmitter type HMP233, abbreviated here as RH air sensors. These sensors were calibrated over a water solution of NaCl, i.e. at 75.36 % RH at 23 °C. The calibration was later re-examined by measuring the RH over a water solution of KNO_3 (94.0 % RH at 23 °C).

The most critical location for condensation was considered to be on the warm side of the VIPs in Field 1 and Field 2 (Fig. 7), and at the warm side of the wind barrier (WB) at the reference field (F3). The thermal resistance of the wind barrier is marginal and the type of wind barrier used has a small water vapour resistance, i.e. condensation was expected to occur on the wind barrier as well as on the VIP surface. Hence, for practical reasons and the desire of having the sensors at the same locations for all four fields, most of the sensors were located at the warm side of the wind barrier. The sensors at the top and the bottom were located 100 mm from the sills where the natural convection in the mineral wool was considered to have the largest influence. The sensors at the middle were located directly within the joints of the VIPs, i.e. 1200 mm above the bottom sill. Furthermore, there were also sensors at other locations in the wall module. A wetness sensor were glued directly over the horizontal joint between two VIPs in F1 and F2 as shown in Fig. 7, i.e. F1 and F2 had wetness sensors on both sides of the wind barrier in the middle of the field. There were also thermocouples in all four ventilated air cavities as depicted in Fig. 7. In addition, air temperature and RH were measured in both climate rooms.

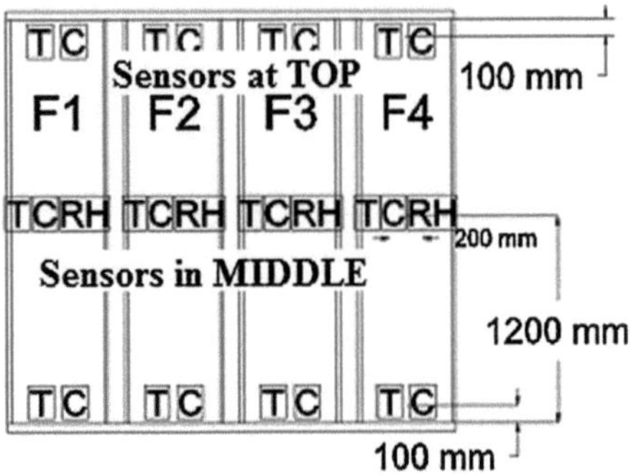

Fig. 8 A section plane of the timber frame module as seen from just within the warm side of the wind barrier. Thermocouples (T), surface wetness sensor (C), relative air humidity (including thermocouple) sensors (RH)

Figure 8 shows the location of the sensors on the wind barrier. The sensors at the top and the bottom are located 100 mm from the sills where the natural convection in the mineral wool is considered to have the largest influence. The sensors at the middle are located directly within the joints of the VIPs, i.e. 1200 mm above the bottom sill.

2.5 Data Logging

The 36 sensors gave a total amount of 42 logging channels due to six combined humidity and temperature sensors (RH air sensors). Three different logging systems were used in the experiment. Due to the large number of sensors, the thermocouples and the air humidity sensors were logged by two systems, i.e. Solartron 35951C I.M.P. connected to Orchestrator 1.4.7 software, and FLUKE Hydra data logger 2625A with appurtenant software. The wetness sensors required a separate logging system, i.e. a logger made by SINTEF containing a Greisinger GMH 830 material moisture meter, which was computer interfaced via LabVIEW8.6 software.

2.6 Test Procedure

The test wall module was built between two climate rooms in the laboratory. The temperature at the cold side was intended to be −20 °C, but due to climate room regulation limitations the outdoor temperature was adjusted to −18 °C. The relative humidity in the outdoor climate was not controlled, but was in the order of RH ~60 %. In the indoor climate the temperature was held constant at 20 °C, while the RH was adjusted in the following steps of 10 %:

1. RH 30 % (~4 g/m^3)
2. RH 40 % (~6 g/m^3)
3. RH 50 % (~8 g/m^3)
4. RH 60 % (~10 g/m^3)

Table 1 Internal humidity classes, derived from buildings in Western Europe (EN ISO 13788 2001)

Humidity class	Internal moisture excess (g/m^3)	Type of building
1	<2]	Warehouse
2	<2–4]	Office, shop
3	<4–6]	Home with few residents
4	<6–8]	Home with many residents, sports centre, kitchen, canteen
5	>8	Special buildings, e.g. laundry, indoor swimming pool

The table relates only to outdoor temperatures below 0 °C, which are the most interesting temperatures

Note that the above given levels in g/m^3 correspond approximately to the internal moisture excess which equals the given indoor RH for an indoor temperature of 20 °C, an outdoor RH of 60 %, and an outdoor temperature of −18 °C.

These moisture levels are henceforth called *climate steps*. The term internal moisture excess represents the difference in moisture content (g/m^3) from the indoor to the outdoor air (EN ISO 13788 2001). To calculate the internal moisture excess the effects of the indoor and outdoor air temperature, RH outside, moisture production inside and ventilation degree are taken into account. The term is useful in practical condensation calculations and divides different building types into humidity classes as seen in Table 1 (EN ISO 13788 2001).

2.7 Numerical Simulations of Temperature and Moisture Performance

Both 1D and 2D simulations of thermal and hygrothermal performance of the fields have been performed. The two simulation programs WUFI Pro (WUFI 1D 2008) and WUFI 2D (WUFI 2D 2010) have been used. The 1D calculations only simulate the gross structure of the fields, i.e. the thermal and hygrothermal performance in the centre of a VIP. The 2D simulations also account for possible heat and vapour transport through the joint between two VIPs. The climate file applied in the simulations is based on the logged temperature and RH in the climate rooms during the experiment. That is, the simulation results and the logged values from the experiment could be compared directly. The material data applied in the simulations was partly from WUFI, partly from EN ISO 10456 (2007) and from some other literature sources. In the simulations, the VIPs were treated as two separate materials in three layers, i.e. VIP envelope, VIP core and again VIP envelope, where the thickness of the VIP envelope was set to 1 mm to avoid problems in the numerical solver. As a result of this, and based on a 0.1 mm thick VIP envelope thermal conductivity of 0.54 W/(mK) (Tenpierik and Cauberg

2007), the adjusted thermal conductivity of the VIP envelope used in the simulations was 5.4 W/(mK) for heat flow perpendicular to the envelope, and 0.054 W/(mK) for heat flow longitudinal to the envelope. As no data was available, the conductivity in these calculations was assumed to be identical in the perpendicular and longitudinal directions. The thermal conductivity of the VIP core was calculated so the total thermal conductivity of the three layers was identical to the declared upper value of 0.005 W/(mK) by the producer (Porextherm 2009). In addition to verify that the simulation programs were reliable for the tested structures, these simulations also represented a control of the measured values in the experiment, especially the wetness sensor measurements. Mainly Field 1 and Field 2 were simulated as they represented the highest risk of moisture problems.

2.8 Condensation Control Calculations

Additional calculations that investigate condensation risk at the VIPs in Field 1 and Field 2 were performed. The basis of the calculations was the standard for hygrothermal performance of building components EN ISO 13788 (2001), which describes a method for addressing a critical limit for surface condensation and accounts for the following parameters:

- Mean monthly external temperature.
- Mean monthly external RH.
- Internal temperature for the respective building.
- Internal moisture excess for the respective building (according to Table 1).

Instead of calculating the critical limit for condensation at one definite climate and internal moisture excess, the method is adapted to calculate the limit where condensation starts for different internal moisture excess and external temperatures. This means that the external RH and the internal temperature were fixed at a probable value (80 % RH and 20 °C). Moisture excess at the condensation limit was calculated for external temperatures from −40 to 20 °C. The calculations were performed for the gross structure, i.e. in the centre of the surface of a VIP. Therefore, a surface resistance of 0.13 (m^2K)/W was used even if the standard sets the worst-case surface resistance to 0.25 (m^2K)/W (EN ISO 13788 2001). The results from the condensation control calculations are shown in the Results section 'Condensation Control Results'.

3 Results and Discussion

3.1 VIP Fastening Bracket

The thermal bridges of different versions of the VIP fastening bracket (made of different materials), of the joint between two VIPs, and of VIPs fastened with a wooden furring strip are depicted in Table 2, which shows some of the calculated linear thermal transmittances that make the basis for the average linear thermal transmittance per meter horizontal joint between two VIPs as calculated in Table 3.

The resulting average linear thermal transmittance per meter of the horizontal joint between two VIPs (Table 3) is as mentioned a result of the fastening bracket thermal transmittance (only present over the wooden stud) and the thermal transmittance of the horizontal joint between two VIPs (from Table 2). Their influence is according to the ratio of their length (e.g. 564 mm of version 1 and 36 mm of version 2 for a section length of 600 mm).

The results (Table 3) show that even if the thermal transmittance of the fastening bracket is larger than the thermal transmittance of the furring strip, the average thermal transmittance per meter of VIP joint is smaller for the fastening bracket. The average linear thermal transmittance of the fastening bracket used in the laboratory experiments (tape-enveloped steel) has been calculated to 0.0083 W/(mK), which is 69 % lower and thus better than the traditional furring strip fastening method with a calculated average linear thermal transmittance of 0.027 W/(mK). Applying tape-enveloped stainless steel (instead of carbon steel) reduces the average linear thermal transmittance further to 0.0067 W/(mK).

3.2 U-Value Calculation Results

The U-values of Field 1 and Field 3 calculated according to EN ISO 6946 (2007) are shown in Table 4. The U-value of Field 1 is calculated with a pristine thermal conductivity of VIP of 0.005 W/(mK) (Porextherm 2009), with a common design conductivity of 0.008 W/(mK) that accounts for ageing of the VIPs, and with a conductivity for punctured VIPs of 0.020 W/(mK) (Binz et al. 2005). The linear thermal transmittance found for the fastening bracket of tape-enveloped steel from Table 3 is used in the U-value calculations. Table 4 shows that 30 mm additional VIPs on the reference field have a large influence on the U-value, even if the VIPs are punctured. The U-value of Field 1 with aged VIPs is at the level of today's requirements in Norway at 0.18 W/(m^2K) (TEK 1997). Field 1 with pristine VIPs however, manages the requirements of the passive house standard in Norway NS 3700 (2010). For how long the VIPs are able to maintain a thermal conductivity low enough to fulfil this demand is not studied within this work. Ageing of VIPs is a subject of investigation and may be found in Wegger et al. (2011). The U-value of Field 2 and Field 4 is not shown here. That is, Field 4 is assumed to have about the same U-

Table 2 Linear thermal transmittance (Ψ) of horizontal joint between two VIPs, the VIP fastening bracket, and a wooden furring strip

Description	U-value/Ψ	Isotherms and colour flux magnitude
Version 1. Thermal bridge of the joint between VIPs including edge effect of VIP envelope and 2 mm air cavity. Vertical cross-section through mineral wool.	U-value = 0.114 $W/(m^2K)$ Ψ = 0.0031 W/(mK)	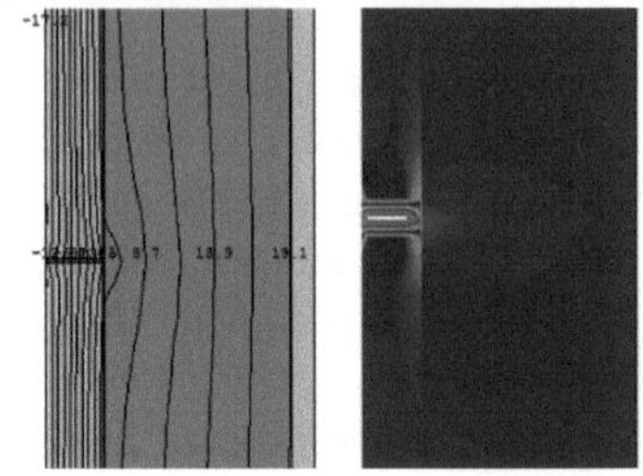
Version 2. Thermal bridge of joint between VIPs including fastening bracket made of steel, edge effect of VIP envelope, and 0.8 mm air cavity. Vertical cross-section through timber frame.	U-value = 0.243 $W/(m^2K)$ Ψ = 0.10 W/(mK)	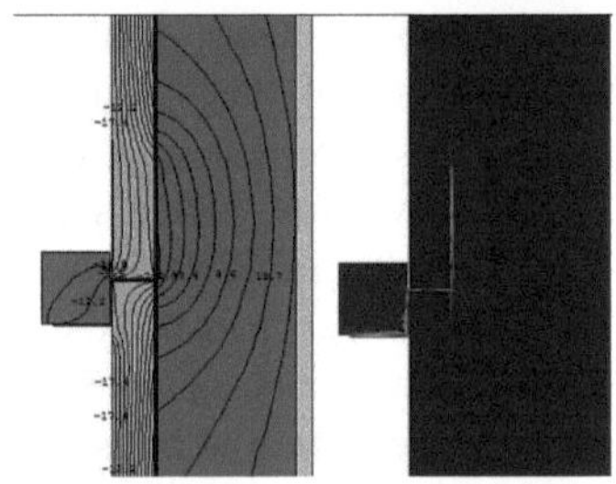
Version 3. Thermal bridge of furring strip between VIPs. Including the wooden furring strip and the edge effect of VIP envelope (no air cavity). Vertical cross-section through mineral wool.	U-value = 0.136 $W/(m^2K)$ Ψ = 0.025 W/(mK)	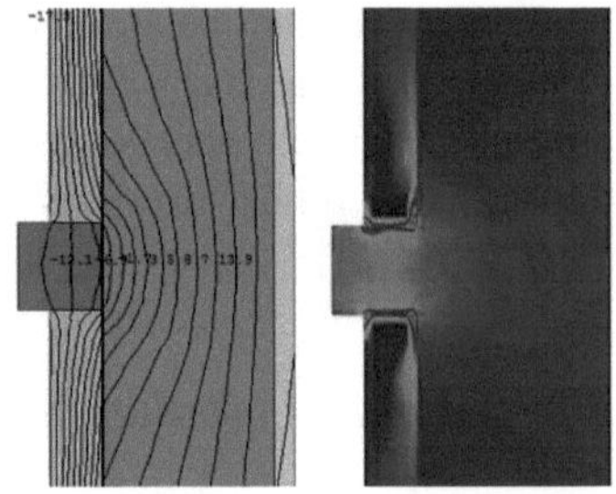
Version 4. Thermal bridge of furring strip between VIPs including the wooden furring strip and the edge effect of VIP envelope (no air cavity). Vertical cross-section through timber frame.	U-value = 0.200 $W/(m^2K)$ Ψ = 0.059 W/(mK)	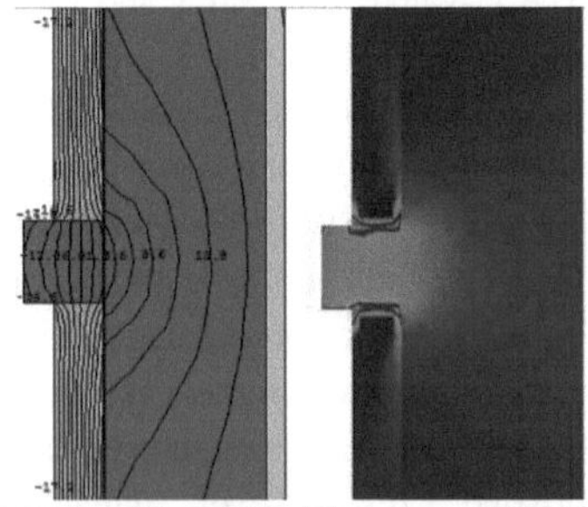

All drawings are shown as vertical cross-sections

value as Field 1. However, some thermal bridges, like the thermal bridge between two floors, will not be thermally insulated when the VIPs are located at the inside. Insulation of

Table 3 Calculation of average linear thermal transmittance per meter joint between two VIPs

Type of fastener	Average linear thermal transmittance (W/(mK))
Fastening bracket made of *steel* $\Psi_{FB\text{-}S}$	0.0091
Fastening bracket made of *tape-enveloped steel* $\Psi_{FB\text{-}TES}$	0.0083
Fastening bracket made of *stainless steel* $\Psi_{FB\text{-}SS}$	0.0077
Fastening bracket made of *tape-enveloped stainless steel* $\Psi_{FB\text{-}TESS}$	0.0067
Furring strip between VIPs $\Psi_{Furring\ Strip}$	0.027

The basis of the calculation is the thermal transmittances from Table 2 and several others not depicted here

Table 4 U-values of Field 1 with different properties of the VIPs and the U-value of Field 3

Field	U-value (W/(m^{2}K))
F1, pristine 30 mm VIPs*	0.143
F1, design (aged) 30 mm VIPs	0.181
F1, punctured 30 mm VIPs	0.262
F3, reference field	0.411
For comparison: Maximum U-value for passive houses in Norway (NS 3700 2010)	0.15

* The small thermal bridge of the vertical VIP joints is not accounted for in the calculations. However, the conductivity of the VIPs is given a conservative pristine value of 0.005 W/(mK) (<0.005 W/(mK) Porextherm 2009)

thermal bridges is a large advantage of adding insulation on the outside instead of the inside of a building structure.

From a thermal insulation point of view, Field 1 is the best of the three retrofitting alternatives evaluated (Fig. 2). The potential moisture problems with this method for improving thermal performance of a wall are addressed in the following sections.

3.3 Wetness Sensor

The tailor-made wetness sensor has proven to be quite trustworthy during the laboratory tests. Even though the sensor was made as an indicator for condensation, the sensor has also proven to be relatively reliable for RH from 70 to 100 %. The wetness sensor and the RH air sensor were in good agreement when placed at the same location, e.g. as in the middle of Field 1 (see Fig. 9). However, the wetness sensor is primarily an indicator of RH and condensation and the reliability

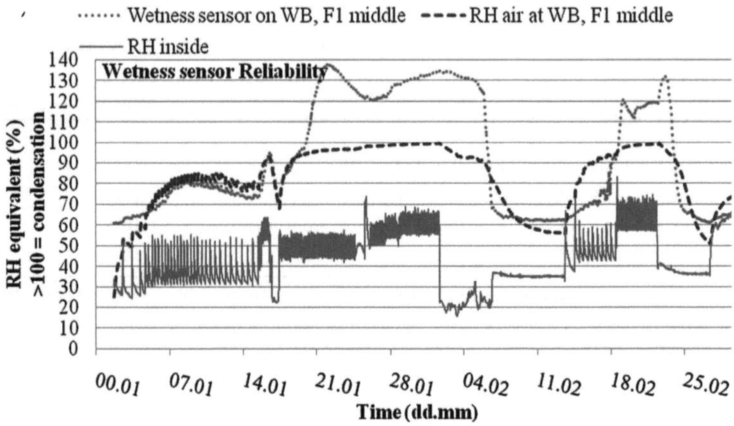

Fig. 9 Testing the tailor-made wetness sensor during two condensation cycles

should be further tested. Note that the wetness sensors show values beyond 100 in the result plots. This is due to the errors concerning the calibration and use of the wood moisture sensor, as mentioned earlier. The values below 100 are equivalent with percent RH while the values above 100 indicate condensation. However, to be sure the value should be above 105 before condensation is denoted. Note that the wetness sensor do not measure lower RH values than about 60 % because the electrical resistance is then too high for the wood moisture meter to measure.

After the experimental testing of the fields was finished, a dehumidifier was placed in the climate room representing inside climate before another cycle of high RH was conducted. As seen in Fig. 9 the wetness sensor seems to be more reliable at its first humidifying. As the wetness sensor and the RH air sensor were more in correspondence at the first humidifying period (from day 01.01 to 21.01) than at the second humidifying (from 11.02 to 27.02), it is possible that the glue on the tape the wetness sensor is made of loosened during the first humidifying cycle. If so, then the contact between the paper and the electrode loosened, i.e. more moisture was required (maybe even free water) before the electrical resistance was lowered significantly.

3.4 Experimental Moisture and Condensation Results

Figures 10, 11, 12 and 13 depict the experimental moisture and condensation results as RH equivalent versus time at different locations inside selected test fields (WB denotes wind barrier), between the following climate steps:

- Climate step 1 and 2 (from 30 to 40 % RH).
- Climate step 2 and 3 (from 40 to 50 % RH).
- Climate step 3 and 4 (from 50 to 60 % RH).

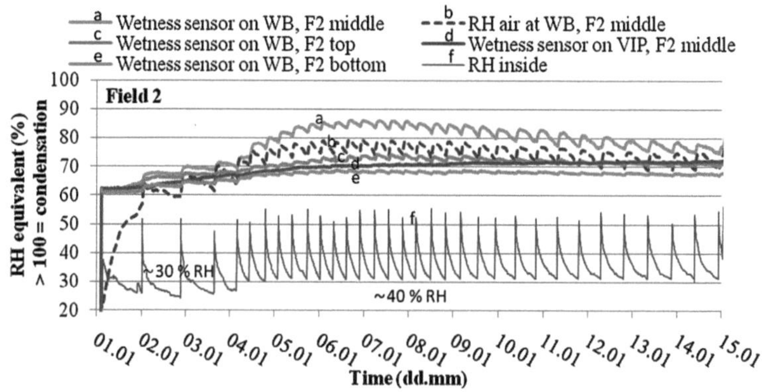

Fig. 10 Moisture sensors in F2 during climate step 1 and 2. The RH equivalent values of the four wetness sensors, as well as the RH air sensor, are shown. The climate room RH was ~30 % from day 1 to 4, and ~40 % from day 4 to 15 (graph f)

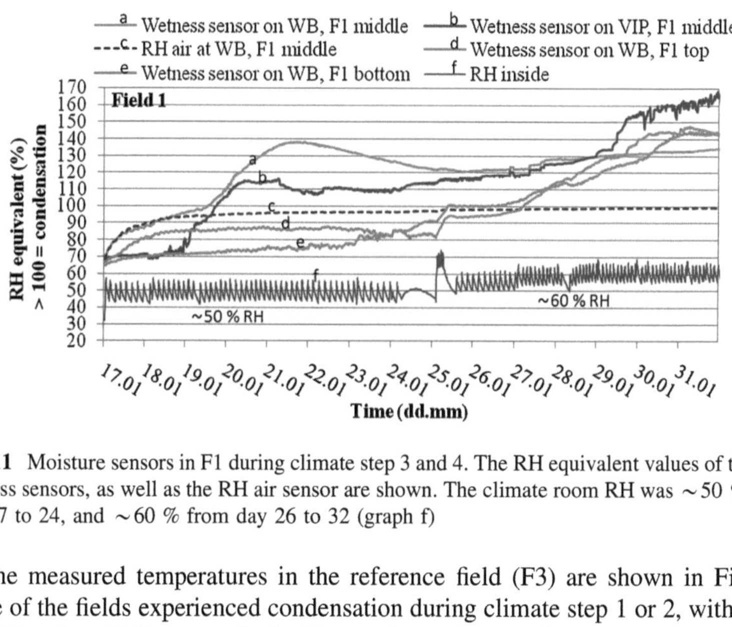

Fig. 11 Moisture sensors in F1 during climate step 3 and 4. The RH equivalent values of the four wetness sensors, as well as the RH air sensor are shown. The climate room RH was ~50 % from day 17 to 24, and ~60 % from day 26 to 32 (graph f)

The measured temperatures in the reference field (F3) are shown in Fig. 14. None of the fields experienced condensation during climate step 1 or 2, with a RH of ~30 and ~40 %, respectively. Hence, day 1–day 15 of the experiment are only plotted for F2, thus depicting a case with no condensation as seen in Fig. 10. The large alternation of RH in the climate room, up to 25 % points from the lowest to the highest value, can be seen in Fig. 10. However, this problem was reduced after test day 15, when a different humidifier was installed.

Figure 11 shows F1, during the two climate steps with RH ~50 and ~60 %. F1 experienced condensation at RH ~50 %, which was not expected for this climate step compared to numerical simulations, at least not after that short period of exposure. This might be caused by the relatively large alternations in RH at the

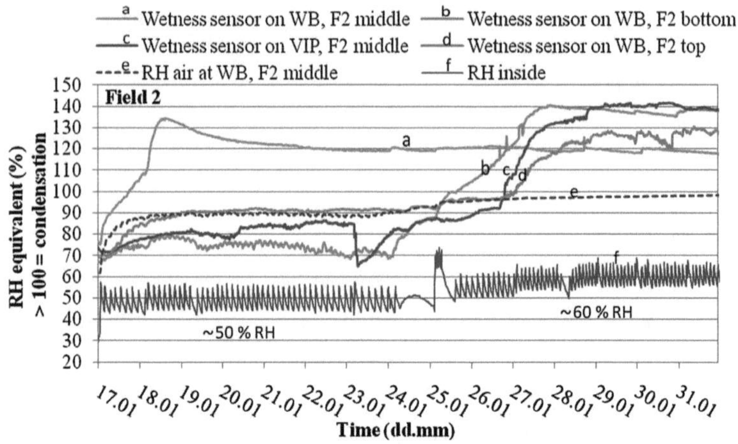

Fig. 12 Moisture sensors in F2 during climate step 3 and 4. The RH equivalent values of the four wetness sensors, as well as the RH air sensor, are shown. The climate room RH was ~50 % from day 17 to 24, and ~60 % from day 26 to 32 (graph f)

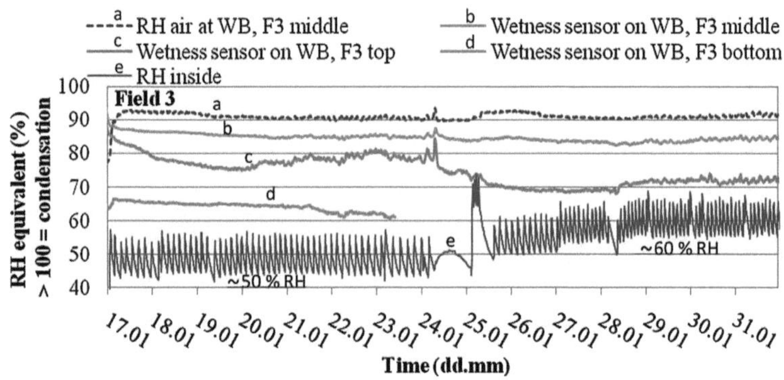

Fig. 13 Moisture sensors in F3 (reference) during climate step 3 and 4. The RH equivalent values of the three wetness sensors, as well as the RH air sensor, are shown. The wetness sensor at the bottom (graph d) stopped giving reliable outputs. The climate room RH was ~50 % from day 17 to 24, and ~60 % from day 26 to 32 (graph e)

inside climate room, seen in the lower graph (graph f) in Fig. 11. The moisturising at the RH peak is possibly larger than the drying-out at the RH bottom. Small air leakages in the structure might also contribute to the measured condensation. The wetness sensor shows condensation (value above 100 %) when the RH air sensor shows about 95 % RH. A possible explanation is that the RH air sensor is located about 5 mm from the wind barrier. Moreover, the wetness sensor is more influenced by condensed water on the vapour barrier than the RH air sensor. The wetness sensor located on the joint between two VIPs shows lower values than the wetness sensor on the wind barrier. Assuming that both sensors have the same temperature, condensation may be caused by the small vapour resistance of the

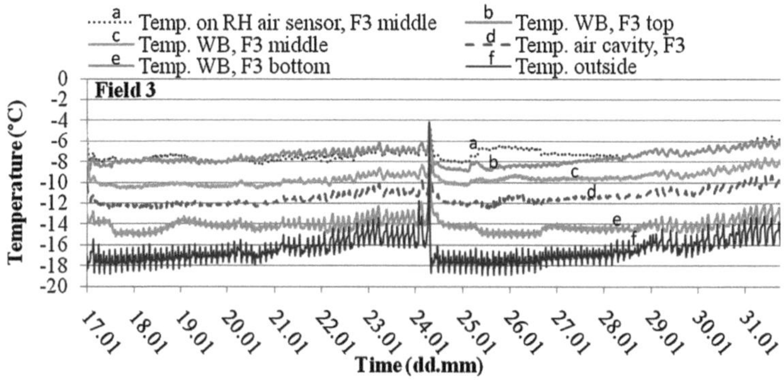

Fig. 14 Temperatures in F3 (reference) measured at the warm side of the wind barrier (WB) over the same time span and at the same locations as the wetness sensors shown in Fig. 13. In addition, the temperature on the RH air sensor, the temperature in the air cavity on the inside of the wooden cladding, and the outside temperature in the climate room, are given

wind barrier. The wetness sensor on the VIP (behind the wind barrier) then becomes moisturised by the wet wind barrier, which causes the sudden increase at the wetness sensor on the VIP. Another influencing factor is convection over the joint between two VIPs (about 2 mm opening) that might dry the sensor somewhat. A third aspect may be time delay due to the small vapour resistance of the wind barrier, but this should not result in such a sudden increase. The wetness sensors at the top and the bottom have different development at the end of climate step 3. A possible explanation might be that the wetness sensor at the top experienced a lower RH caused by an (erroneous) increase in the outdoor temperature during the experiment (Fig. 14). The increase at the wetness sensor at the bottom may be condensed water that runs down from the middle of the vapour barrier and moisturises the sensor.

Figure 12 shows F2, during the two climate steps with RH ~50 and ~60 %. In accordance with numerical simulations, the field experienced condensation during both climate step 3 and climate step 4. However, an interesting difference between F1 and F2 is the wetness sensors located on the VIPs. In F1 this sensor measured condensation at both climate steps, whereas in F2 this sensor measured condensation only for climate step 4 even if this field is less insulated (only 20 mm VIP). The temperatures measured at the warm side of the wind barrier (Fig. 14) depict that the wetness sensor at F1 had a higher temperature than the wetness sensor in F2, i.e. as expected, thus the reason why the wetness sensor at the joint between two VIPs in F2 did not show condensation is unclear. It might be caused by difference in workmanship of the wall or installation of the sensor. The somewhat rigid wind barrier was hard to fit properly and this might have caused air cavities between the VIP and the wind barrier that was different from F1 to F2. Another unexpected occurrence was the sudden drop at the wetness sensor at the joint between two VIPs (graph c) in Fig. 12, which might be caused by a loosening of the sensor from the surface of the VIPs (however, this is not known).

Both Figs. 11 and 12 show a peak at the wetness sensors during condensation. The cause of these peaks might be the different sorption curves for spruce and paper, i.e. that the measured electrical resistance is in a sensitive area of the curve that describes the electrical resistance as a function of moisture content for spruce (as the moisture meter is calibrated with respect to spruce). A small decrease in RH might then give a large effect on the output value of the moisture meter. The graphical plots from F1 and F2 show a lower RH at the top and the bottom than in the middle of the fields. The high RH at the middle is due to the thermal bridge of the joint between two VIPs. The temperature at the structures laying above and below the module influences the temperature, and hence the RH at the top and the bottom. However, the surrounding structures cool down the module, therefore this only has a conservative effect, i.e. lower temperatures increase the RH. It is therefore clear that the thermal bridge at the joint between two VIPs is a critical point for condensation in the wall, together with the surfaces of the fastening brackets.

Field 4 with interior VIP retrofitting performed well during the entire test, i.e. no condensation occurred. RH at the middle of the wind barrier was about 50 % during the whole experiment, thus far from the risk of condensation. The results from F4 are thus in this context not found necessary to depict in graphical plots.

Figure 13 shows the reference field (F3) during climate step 3 and climate step 4. None of the sensors show condensation. The wetness sensors are not calibrated for temperatures below 0 °C and are thus less trustworthy in F3. It is reasonable that the electrical resistance increases when the temperature drops below 0 °C and the water freezes. Therefore, the values from the wetness sensors in F3 are lower than the RH air sensor. Condensation might be present in F3 without being measured. The time frame of the test is not necessarily long enough to give condensation, as F3 will experience a certain exsiccation to the outside climate room. The outside temperature was in addition increasing somewhat from the start of the climate step to the end as the cooling unit was manually de-iced for each climate step (Fig. 14). This contributed to hold the RH air sensor on the wind barrier stable, as seen in Fig. 13.

The results from the experiments allow for use of 30 mm VIPs at the outside of a 100 mm mineral wool wall to improve thermal insulation. However, the method of improving thermal insulation provide certain limitations to the interior and exterior climate. That is, not more severe than 20 °C and an internal moisture excess of maximum 6 g/m^3 at the interior side, and −18 °C and RH 60 % at exterior side. It must be emphasized that these results are only valid for VIPs in a pristine condition and when designing a real building the aged condition of the VIPs must be taken into account.

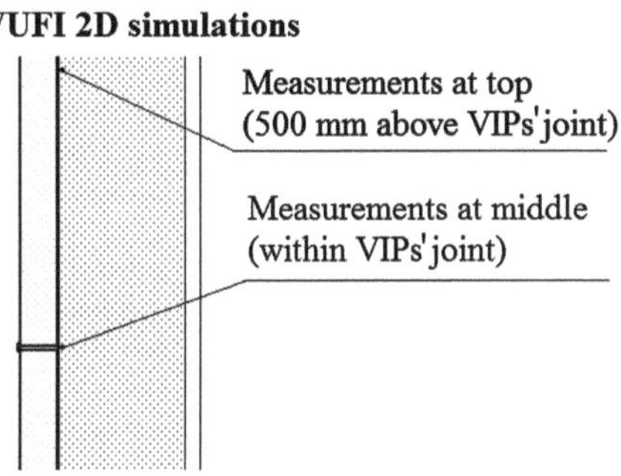

Fig. 15 Illustration of measurement points in the WUFI 2D simulations

3.5 Comparison of Measured and Simulated Results

An illustration of the WUFI 2D simulations is shown in Fig. 15. The measurement points are located at the warm side of the wind barrier, one within the joint between two VIPs and one 500 mm above the joint between two VIPs.

The WUFI simulation results are shown in Figs. 16, 17, 18, where measured (logged) values from the wetness sensors and the RH air sensors are plotted for comparison. These graphical plots show only the RH equivalent part of the wetness sensors, i.e. values above 100 are plotted as 100 % RH.

The WUFI and the wetness sensors results at the top of F2 are depicted in Fig. 16. The difference in RH values between the 1D and 2D simulation seems to be systematic and is probably caused by a lower temperature in the 2D simulation as depicted in Fig. 17, which may be caused by the thermal bridge in the 2D simulation. The difference in wetness sensor measurements and simulated RH values seems mainly to be caused by the temperature difference as well. In Fig. 17 the logged temperature graph is about 2.5 °C warmer than the simulated values, which corresponds to a decrease in RH of about 20 % points at this temperature level. The low temperatures in the simulations may be caused by the application of the VIP upper limit declared thermal conductivity of 0.005 W/(mK) (Porextherm 2009), as the real thermal conductivity may be as low as 0.004 W/(mK) (Binz et al. 2005). Measurements have shown that both the conductivity and the thickness of different VIPs may vary, e.g. Grynning et al. (2011) measured thermal conductivities somewhat higher than the declared ones and panel thicknesses that were less than stated.

Furthermore, analogous plots (not presented here) as given in Figs. 16 and 17 were also made for F1, i.e. simulated and measured values of RH and temperature at the top of F1. The WUFI 1D and WUFI 2D simulation results compared well. However, the graph of the wetness sensor at the top showed also in this plot about 10 % points lower RH, except for the case of condensation. The difference seemed also here to be caused by a warmer temperature in the experimental results.

Figure 18 depicts measured and simulated RH values at the middle of F1, where both the simulated and measured values compared quite well. The wetness sensor showed condensation when the two other graphs showed a RH of about

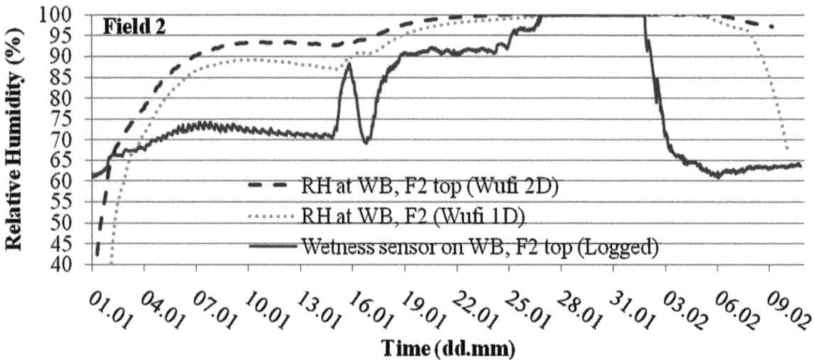

Fig. 16 RH simulated in WUFI 1D and WUFI 2D, and measured RH at the top of F2

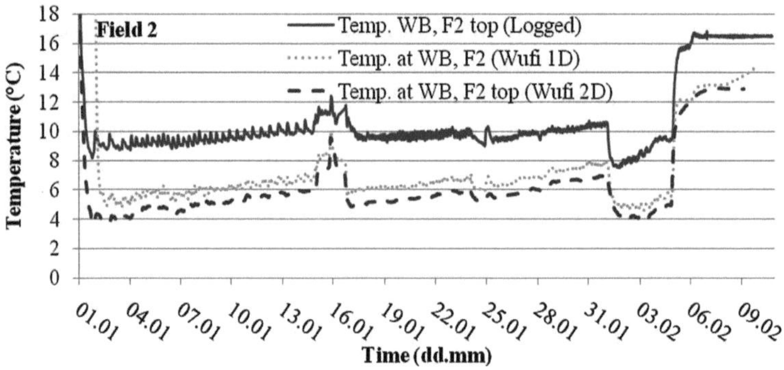

Fig. 17 Temperature simulated in WUFI 1D and WUFI 2D, and measured temperature at the top of F2

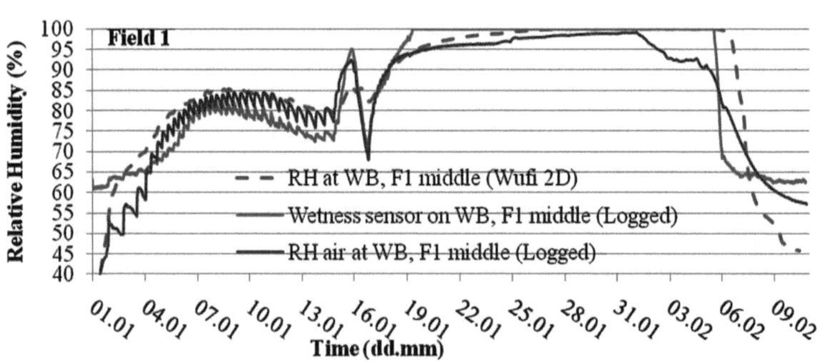

Fig. 18 RH simulated in WUFI 2D, and measured RH in the middle of F1

95 %, thus indicating that the wetness sensor represents a better condensation sensor than the regular RH air sensor. That is, the wetness sensor has a better ability to register wetness on a surface. The small differences might be due to the eccentric location of the two other measurements, i.e. they are placed on a slightly warmer isotherm.

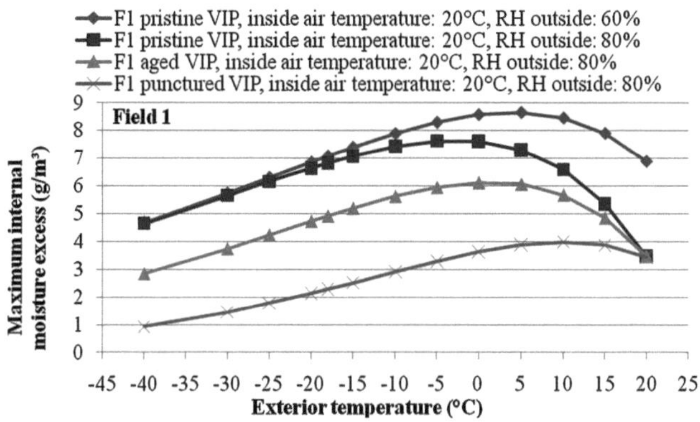

Fig. 19 Maximum moisture excess before condensation on VIP in F1

The effect of the drying-out process at the joint between two VIPs was questioned prior to the experiments. This mechanism is not numerically simulated, but the results have shown that the reduced temperature at this location has a larger influence on the RH than the drying-out effect, i.e. the RH at the joint between two VIPs was higher than in the centre of a VIP (sensors at middle compared to sensors at top).

Generally, there is good agreement between the measured and simulated results. The differences registered seem mainly to be caused by inaccurate material properties applied in the simulations. These results show that both the WUFI simulations and the experimental results are reliable, and strengthen the reliability of the wetness sensor as a RH indicator.

3.6 Condensation Control Results

Figures 19 and 20 depict the calculations of condensation-start-limits, illustrating what external temperature and internal moisture excess may cause condensation on the VIP surface for different conditions (thermal conductivities) of the VIPs. The calculations are based on an interior temperature of 20 °C and an exterior RH of 80 % (except the upper graphs with an exterior RH of 60 %). As for the U-value calculations, the thermal conductivities applied for pristine, aged and punctured VIPs are employed. It is important to emphasize that the tested wall constructions do not have a vapour barrier on the inside, i.e. considered as a worst-case wall compared to a more typical wall with a vapour barrier. The water vapour diffusion resistance of the materials at the interior of the VIPs is neglected, which represents a conservative simplification. It should also be noted that the risk of mould growth on wood in structures normally is assumed to start at a RH of 80 %.

Figure 19 shows that for an exterior RH level of 60 % and an exterior temperature of −18 °C, the moisture excess must be 7.1 g/m^3 for condensation to occur at the VIPs in F1. This is equivalent with an indoor RH of 46 % at these

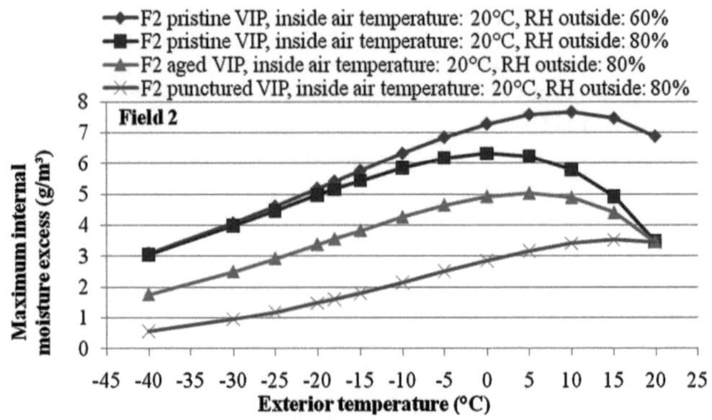

Fig. 20 Maximum moisture excess before condensation on VIP in F2

conditions. During the experiment, condensation was not observed on the top or bottom of the fields before the inside RH was 60 %. Neglecting the water vapour diffusion resistance of the materials on the interior side of the VIPs may explain this difference. The material thermal conductivity might also deviate from the real value and cause an incorrect temperature factor. Lower thermal resistance at the surface due to air circulation in the climate rooms may also contribute to this difference. Furthermore, the calculations assume steady-state conditions, whereas the experiment was not run for a sufficient long time to reach complete steady-state.

The exterior temperature is to be the minimum mean temperature over three days in condensation calculations (SINTEF 471.111, 1999), and in design evaluations the aged value of the thermal conductivity of VIPs should be applied. Applying this for Oslo, Norway, with a mean minimum temperature over three days of −19.8 °C (SINTEF 451.021 2009), would give a maximum internal moisture excess of 4.7 g/m^3 (assuming an internal temperature of 20 °C or higher). Note that the external RH at such low temperatures does not make a large difference to the calculations. Field 1 performs badly both thermally and hygrothermally if the VIPs are punctured.

Figure 20 shows the condensation limits for F2, which is less insulated than F1 and hence cannot handle that large moisture excess before condensation occurs on the VIPs. Compared to F1 in Fig. 19 the graphs for F2 in Fig. 20 are roughly shifted 1 g/m^3 lower. Field 2 might still perform well enough at many locations given a low moisture excess or a temperature within a specific range. Installing a ventilation system that prevents high indoor RH at wintertime, may also be a way of avoiding condensation in the wall.

These calculations only apply for the gross structure (i.e. at the centre of a VIP surface) and the temperature factor will be lower at the thermal bridges in the structure. However, this is normally only at small areas. A tight vapour barrier at the inside will avoid the condensation problem. Still, it is important to address these potential condensation problems without a vapour barrier as many are sceptical to adding a vapour tight material at the exterior side of an insulated structure. A non-tight vapour barrier may lead to accumulation of moisture between two vapour tight materials.

An increase in outdoor RH makes little difference at low exterior temperatures, but large difference when the exterior temperature is approaching the interior temperature. However, in Norway the relevance of the calculations is less for exterior temperatures above about 10 °C, since the exterior RH is less likely to be as high as 80 % the higher the temperature is. In addition, the residents of a building start to ventilate way more at warm exterior temperatures, e.g. opening windows, etc., and thus, the air change rate increases and consequently the moisture excess is lowered. Furthermore, it is important to note that if a lower interior temperature (here <20 °C) had been used, the accepted moisture excess would be reduced, i.e. an increased risk of condensation and hence less moisture robustness.

4 Conclusions

Experimental and theoretical investigations have been carried out on a test module built between two climate rooms with an indoor and an outdoor climate, respectively. The module consists of four wall fields, one reference field representing a timber frame wall built according to building regulations from the 1970s in Norway, and three fields representing different ways of improving the thermal insulation of the reference field by retrofitting with vacuum insulation panels (VIPs). Two of the fields were built with VIPs on the exterior side of the timber frame wall, one field with 20 mm VIPs and one field with 30 mm VIPs. The fourth field was built with VIPs retrofitted on the interior side. Temperature, relative humidity (RH) and surface wetness were measured during the experiment, as vapour tight VIPs on the exterior side may introduce condensation in a wall. Numerical simulations for the wall fields were carried out and compared to the experiments. The simulations showed good correlation with the experimental results. Calculations of threshold limits for condensation on the VIPs in the wall structures were performed with respect to different outdoor temperatures and internal moisture excess.

For application in the retrofitting experiments, a tailor-made fastening bracket for VIPs was developed, which reduced the thermal bridge of the joint between two VIPs compared with regular methods of fastening VIPs, i.e. using furring strips. Furthermore, a tailor-made moisture or wetness sensor was fabricated and applied in the experiments, which enabled moisture and condensation measurements directly on various surfaces. The wetness sensor proved to be relatively reliable for RH levels between 70–100 % and for indicating condensation. However, the sensor does not perform at temperatures below the freezing point, at RH levels below 60 %, or for repeating cycles of high and low RH. The reliability of the wetness sensor was strengthened by the results from the numerical simulations.

The results from the experiments, the simulations, and the condensation controls conclude that timber frame buildings thermally insulated with 100 mm mineral wool, might be retrofitted at the exterior side by adding 30 mm VIPs in a continuous layer if certain limits to outdoor temperature, internal moisture excess and indoor temperature are provided. The condensation calculations emphasized the importance of avoiding puncturing of VIPs and in addition to account for the aged conditions of VIPs. Hence, it is strongly emphasized that any potential or likely changes in the VIP thermal conductivity have to be accounted for during the design of retrofitting projects, where VIPs are applied on the exterior side of a wall.

Acknowledgements This work has been supported by the Research Council of Norway and several partners through the SINTEF and NTNU research projects "*Robust Envelope Construction Details for Buildings of the 21st Century*" (ROBUST) and "*The Research Centre on Zero Emission Buildings*" (ZEB). Franco Bløchlinger from Metallplan and the manufacturer Porextherm is acknowledged for supplying the vacuum insulation panel test samples. Ole Aunrønning (NTNU) and Egil Rognvik (SINTEF) provided valuable help during various experimental tasks.

References

Al-Homoud, M.S.: Performance characteristics and practical applications of common building thermal insulation materials. Build. Environ. **40**, 353–366 (2005)

Baetens, R., Jelle, B.P., Thue, J.V., Tenpierik, M.J., Grynning, S., Uvsløkk, S., Gustavsen, A.: Vacuum insulation panels for building applications: a review and beyond. Energy Build. **42**, 147–172 (2010a)

Baetens, R., Jelle, B.P., Gustavsen, A.: Phase change materials for building applications: a state-of-the-art review. Energy Build. **42**, 1361–1368 (2010b)

Baetens, R., Jelle, B.P., Gustavsen, A., Grynning, S.: Gas-filled panels for building applications: a state-of-the-art review. Energy Build. **42**, 1969–1975 (2010c)

Baetens, R., Jelle, B.P., Gustavsen, A.: Aerogel insulation for building applications: a state-of-the-art review. Energy Build. **43**, 761–769 (2011)

Binz, A., Moosmann, A., Steinke, G., Schonhardt, U., Fregnan, F., Simmler, H., Brunner, S., Ghazi, K., Bundi, R., Heinemann, U., Schwab, H., Cauberg, J.M., Tenpierik, M.J., Johannesson, G., Thorsell, T., Erb, M., Nussbaumer, B.: Vacuum insulation in the building sector. Systems and applications (Subtask B). Final report for the IEA/ECBCS Annex 39 HiPTI-project (2005)

Brunner, S., Gasser, P., Simmler, H., Ghazi Wakili, K.: Investigation of multilayered aluminium-coated polymer laminates by focused ion beam (FIB) etching. Surf. Coat. Technol. **200**, 5908–5914 (2006)

Building regulations: Kap. 54–Varmeisolering og tetthet (Norwegian building regulations of 1969, chapter 54–Thermal insulation and air tightness) (1969)

Demirbas, M.F.: Thermal energy storage and phase change materials: an overview. Energy Sources Part B **1**, 85–95 (2006)

EN ISO 6946: Building components and building elements—thermal resistance and thermal transmittance—calculation method (2007)

EN ISO 10211: Thermal bridges in building construction—heat flows and surface temperatures—detailed calculations (2007)

EN ISO 10456: Building materials and products—hygrothermal properties—tabulated design values and procedures for determining declared and design thermal values (2007)

EN ISO 13788: Hygrothermal performance of building components and building elements—internal surface temperature to avoid critical surface humidity and interstitial condensation—Calculation methods (2001)

Farid, M.M., Khudhair, A.M., Razack, S.A.K., Al–Hallaj, S.: A review on phase change energy storage: materials and applications. Energy Convers. Manag. **45**, 1597–1615 (2004)

Geving, S., Erichsen, T.H., Nore, K., Time, B.: Hygrothermal conditions in wooden claddings—test house measurements. Report from the R&D program "Climate 2000", Project report no. 407, Norwegian Building Research Institute, Norway (2006)

Griffith, B.T., Türler, D., Arashteh, D.: Optimizing the effective conductivity and cost of gas–filled panel thermal insulations. In: Proceedings of the 22nd International Thermal Conductivity Conference, Arizona State University, 7–10 November 1993

Griffith, B.T., Arashteh, D., Türler, D.: Gas–filled panels: an update on applications in the building thermal envelope. In: Proceedings of the BETEC Fall Symposium, Superinsulations and the Building Envelope, Washington, DC, 14 November 1995

Grynning, S., Jelle, B.P., Uvsløkk, S., Gustavsen, A., Baetens, R., Caps, R., Meløysund, V.: Hot box investigations and theoretical assessments of miscellaneous vacuum insulation panel configurations in building envelopes. J. Build. Phys. **34**, 297–324 (2011)

Hasnain, S.M.: Review on sustainable thermal energy storage technologies, part I: heat storage materials and techniques. Energy Convers. Manag. **39**, 1127–1138 (1998)

Hostler, S.R., Abramson, A.R., Gawryla, M.D., Bandi, S.A., Schiraldi, D.A.: Thermal conductivity of a clay-based aerogel. Int. J. Heat Mass Transf. **52**, 665–669 (2008)

Jelle, B.P., Gustavsen, A., Baetens, R.: The path to the high performance thermal building insulation materials and solutions of tomorrow. J. Build. Phys. **34**, 99–123 (2010)

Jelle, B.P.: Traditional, state-of-the-art and future thermal building insulation materials and solutions—properties, requirements and possibilities. Energy Build. **43**, 2549–2563 (2011)

Kneifel, J.: Life-cycle carbon and cost analysis of energy efficiency measures in new commercial buildings. Energy Build. **42**, 333–340 (2010)

McKinsey: Pathways to a low-carbon economy. Version 2 of global greenhouse gas abatement cost curve, McKinsey & Company (2009)

NS 3700: Kriterier for passivhus og lavenergihus. Boligbygninger (Criteria for passive houses and low energy houses. Residential buildings) (2010)

Papadopoulos, A.M.: State of the art in thermal insulation materials and aims for future developments. Energy Build. **37**, 77–86 (2005)

Porextherm: Vacupor NT—B2, data sheet. http://www.porextherm.com/web/pdfdownload.htm?ID=255&doctype=PDF_EN&lang=en&externeswindow=2&ts=1270722658 (2009). Accessed 8 April 2010

Schultz, J.M., Jensen, K.I., Kristiansen, F.H.: Super insulating aerogel glazing. Sol. Energy Mater. Sol. Cells **89**, 275–285 (2005)

Schultz, J.M., Jensen, K.I.: Evacuated aerogel glazings. Vacuum **82**, 723–729 (2008)

SINTEF 451.021: SINTEF building research design guides no. 451.021. Klimadata for termisk dimensjonering og frostsikring (Climate data for thermal design and frost control), SINTEF Building and Infrastructure, Norway (2009)

SINTEF 471.111: SINTEF building research design guides no. 471.111. Beregningsmetode for å unngå kondens eller muggvekst på overflater (Calculation method for avoiding condensation or mould growth at surfaces), SINTEF Building and Infrastructure, Norway (1999)

SINTEF 523.002: SINTEF Building Research Design Guides no. 523.002. Yttervegger over terreng. Egenskaper og konstruksjonsprinsipper. Krav og anbefalinger (Building walls above ground. Properties and construction principles. Requirements and recommendations), SINTEF Building and Infrastructure, Norway (2008)

Statistics Norway: Boliger, etter bygningstype og byggeår (K) (Buildings, ranged by building type and construction year), Table 06266 (2009)

Sveipe, E., Jelle, B.P., Wegger, E., Uvsløkk S., Thue, J.V., Grynning, S., Gustavsen, A.: Retrofitting of timber frame walls by application of vacuum insulation panels and investigations of moisture robustness. In: Proceedings of the 12th International Conference on Durability of Building Materials and Components, pp. 571–578, Porto, Portugal, 12–15 April (2011a)

Sveipe, E., Jelle, B.P., Wegger, E., Uvsløkk, S., Grynning, S., Thue, J.V., Time, B., Gustavsen, A.: Improving thermal insulation of timber frame walls by retrofitting with vacuum insulation panels—experimental and theoretical investigations. J. Build. Phys. **35**, 168–188 (2011b)

TEK, §8-2: Energibruk (energikrav revidert i 2007) (Building regulations of 1997, energy demands corrected in 2007, §8-2, Energy) (1997)

Tenpierik, M.J., Cauberg, H.: Analytical models for calculating thermal bridge effects caused by thin high barrier envelopes around vacuum insulation panels. J. Build. Phys. **30**, 185–215 (2007)

Tenpierik, M.: Vacuum insulation panels applied in building constructions. Ph.D thesis, Technische Universiteit Delft, Netherlands (2009)

THERM: Finite element simulator version 5.2.14 (2003)

Wegger, E., Jelle, B.P., Sveipe, E., Grynning, S., Gustavsen, A., Baetens, R., Thue, J.V.: Aging effects on thermal properties and service life of vacuum insulation panels. J. Build. Phys. **35**, 128–167 (2011)

Willems, M.K., Schild, K., Hellinger, G.: Numerical investigation on thermal bridge effects in vacuum insulating elements. In: Proceedings of 7th International Vacuum Insulation Symposium, EMPA, pp. 145–152 (2005)

WUFI 1D: WUFI Pro, Version 4.2 NBI/NTNU. Release 4.2.0.421.DB.24.63 (2008)

WUFI 2D: Version 3.3, Release 3.3.1.60.DB.24.63 (2010)

Wind Resistance Evaluation to Dynamic Response of Mechanically Anchored Waterproofing Membrane System

Hiroyuki Miyauchi, Bartko Michal, Nobuo Katou and Kyoji Tanaka

Abstract Damages of mechanically-anchored waterproofing membrane system from strong wind have been occurring. Thus, this study provides investigation on the behavior of waterproofing sheets and fasteners exposed during typhoon, and the difference between the static pressure test and typhoon test. The polyvinyl chloride waterproofing membrane reinforced with polyester fiber was constructed on a flat roof with dimensions of 6 m by 6 m on the island of Miyako in Japan. At first, a pressure test was carried out by using a compressor. Afterwards, the response of waterproofing membrane system located atop a building when subjected to strong winds from a typhoon in 2009 was measured. Both axial force and lateral force of mechanically-anchored waterproofing membrane system were measured under high speed wind, and lateral force was significantly large. It is presumed that such lateral force causes waterproof membrane damage that ripped the sheet in the lateral direction as if in a typhoon. Next, this study provided the proposal of simple wind resistance test method for mechanically anchored waterproofing membrane system. At first, the pressure test on the flat roof of actual building was carried out to obtain basic data for wind resistance design of this system. As a result, Fastener's force of waterproofing membrane was equivalent to

H. Miyauchi (✉) · B. Michal
Chungnam National University, Department of Architectural Engineering, Daejeon, South Korea
e-mail: miyauchi@cnu.ac.kr

B. Michal
e-mail: bartko@cnu.ac.kr

N. Katou
Tokyu Construction, Institute of Technology, Kanagawa, Japan
e-mail: katou.nobuo@tokyu-cnst.co.jp

K. Tanaka
Tokyo Institute of Technology, Kanagawa, Japan
e-mail: tanaka-kyoji@kme.biglobe.ne.jp

V. P. de Freitas and J. M. P. Q. Delgado (eds.), *Hygrothermal Behavior, Building Pathology and Durability*, Building Pathology and Rehabilitation 1,
DOI: 10.1007/978-3-642-31158-1_10, © Springer-Verlag Berlin Heidelberg 2013

the wind force. Then, the ring type specimen with 200 and 600 mm diameter was made, and the fatigue coefficient of this system after using 10 years was calculated by using the tension test and the SIGDERS test. Finally, the design method for fastener based on fatigue model was proposed in order to apply to construction system.

Keywords Waterproofing membrane · Wind resistance · Field test · Test method · Fatigue

1 Introduction

Mechanically anchored waterproofing membrane systems are directly fixed to the roof substrate with a series of fasteners. Comparing the performance of these systems to that of an adhered waterproofing membrane system, mechanically anchored waterproofing membrane systems do not depend on the conditions of the substrate to ensure adequate attachment to the roof and are also reasonably easy to replace, given the means of attachment. However, the wind resistance of these membrane systems depends on the reliability of the fasteners, and there is some apprehension as to the safety of these systems in respect to wind resistance, especially when they are subjected to high winds loads, such as those occurring during a typhoon. For example, the maximum instantaneous wind speeds exceeded 50 m/s (180 kph) for typhoons No.18 and No.22 that recently occurred in Japan, and this apprehensiveness may be well founded.

Recently, a method for testing the wind resistance of waterproofing membrane systems was proposed and tested in Canada by the Special Interest Group for Dynamic Evaluation of Roofing Systems (SIGDERS) (Baskaran et al. 2009, 2008, 2007, 2005, 1999; Lei et al. 2001). A cycle model of wind loads method was proposed by Uematsu (Uematsu et al. 2009; Furuichi et al. 2005) in Japan for buildings that incorporate a parapet. The present authors have tested the behavior of a mechanically anchored waterproofing membrane system in conjunction with an actual size test specimen by using a fatigue test apparatus (Bartko et al. 2006, 2005) and wind tunnel tests [Ichikawa et al. 2004, 2007]. However, the obtained wind load data are based on wind tunnel tests and laboratory tests, and the relationship between the data and the actual outdoor environment is not clear. Therefore, we constructed a test building on the island of Miyako, in Japan, and installed a mechanically anchored waterproofing membrane system on the roof.

In this study, the static pressure test and the typhoon test were carried out on a flat roof of the outdoor test building. Afterwards, the lateral and vertical force-induced fasteners around the waterproofing membrane were measured by a 6-component force transducer during a typhoon in 2009, and both wind pressure and force were compared.

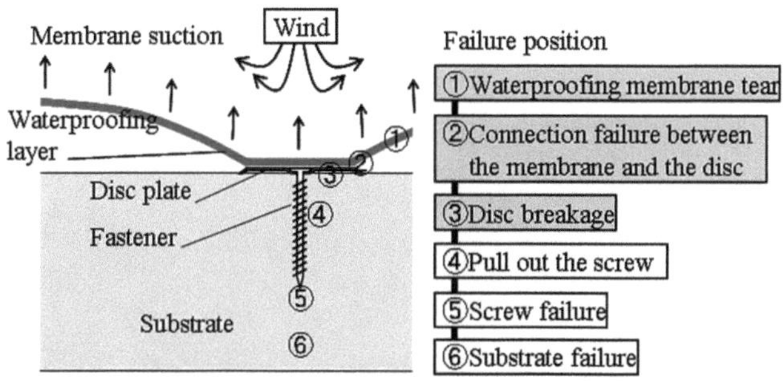

Fig. 1 Wind effects on assemblies in mechanically attached waterproofing membrane

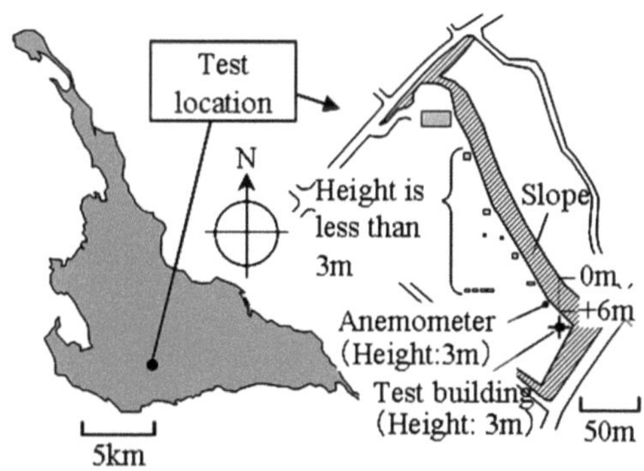

Fig. 2 Test location in Miyako Island

Then, from the results of tensile test and SIGDERS test fatigue factor was calculated. Calculation method of needed number of fasteners based on the cumulative fatigue was presented. And, in this research based on the previous experiment results history (Bartko et al. 2006, 2005 as shown in Fig. 1, we focused on the position of the tear of the waterproofing membrane, failure of a disc as well as of failure of an adhesive.

2 Outline of Measurement in Field Test

2.1 Test Location and Test Building

The measurements in the field test were performed at the Miyako Island test site of the Japan Weathering Test Center, in the southern part of Miyako Island City in Okinawa Prefecture (Fig. 2). Sloping land is at the north face of the test building and is 6 m lower than the ground where the test building is built. However, the area surrounding the building is generally flat and there are no obstructions to the flow of wind across the test building. The building constructed for the tests is shown in Fig. 3. The test building is a rectangular structure made of reinforced

Fig. 3 Test building

Fig. 4 Mechanical property of PVC sheet (Longitudinal direction)

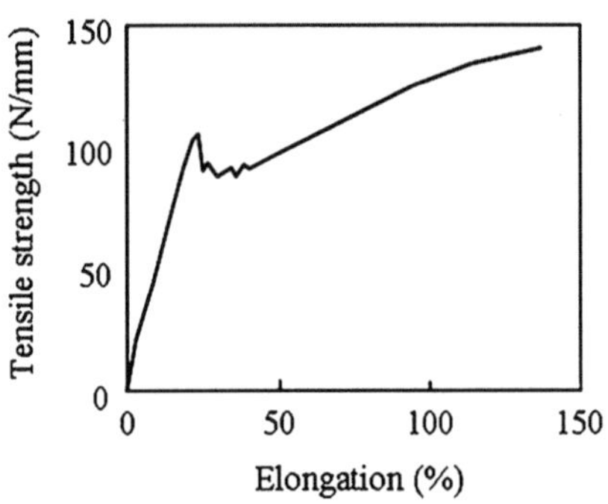

concrete with dimensions of 6 m by 6 m width and 3 m height, and incorporates a waterproofing membrane on a flat roof without any parapets.

2.2 Waterproofing Membrane System

The 1.5-mm thick waterproofing membrane is made of polyvinyl chloride (PVC) reinforced with polyester fiber. The polyester fibers, having a thickness of 556 dtex, are mixed at a density of 1.5filaments/cm in both directions of the sheet. The relationship between the elongation and the tensile stress of a dumbbell shaped PVC sheet is shown in Fig. 4.

The waterproofing membrane was fixed using a prefabricated technique, whereby the membrane is mechanically fastened using circular fixing plates. After drilling holes into the concrete substrate, as shown in the upper part of Fig. 5 plates (ext. diam.: 75 mm; thickness: 1.7 mm) were fixed by means of anchors (stainless steel screws of diam. 6 mm and nylon plugs) inserted into the drilled holes and

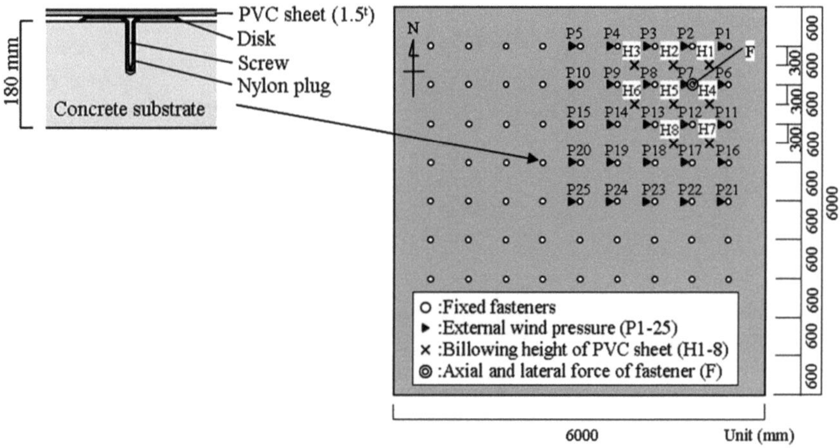

Fig. 5 Waterproofing membrane system and outline of the measurement

non-styrene epoxy acrylate resin adhesive. The waterproofing membrane was then adhered to the plates. Fasteners were spaced 0.6 m apart (a typical value), as shown in Fig. 5, resulting in a total of 81 points over the surface of the roof.

3 Static Pressure Test Using Test Building Roof

3.1 Outline of Pressure Test Method

The static pressure test method is shown in Fig. 6. 40-mm holes in the concrete substrate were drilled to apply pressurized air from the compressor to the space between the waterproofing membrane and the concrete substrate, and the pressure hole was connected with the pressure entrance of the compressor. Pressure was then applied from the back of the waterproof membrane, and the waterproof membrane was billowed artificially. The pressure between the waterproofing membrane and the concrete substrate were measured by using a pressure-tube anemometer through a vinyl tube pipe. The measurement parameters are the pressure between the waterproofing membrane and concrete substrate, the billowing height of the waterproofing membrane, and the force induced in the fasteners.

3.2 Test Results

The billowing condition of the waterproofing membrane during the static pressure test is shown in Fig. 7. The waterproofing membrane was pushed up by the pressure applied by the compressor, and it was assumed that the stress was concentrated in the waterproofing membrane around the fasteners.

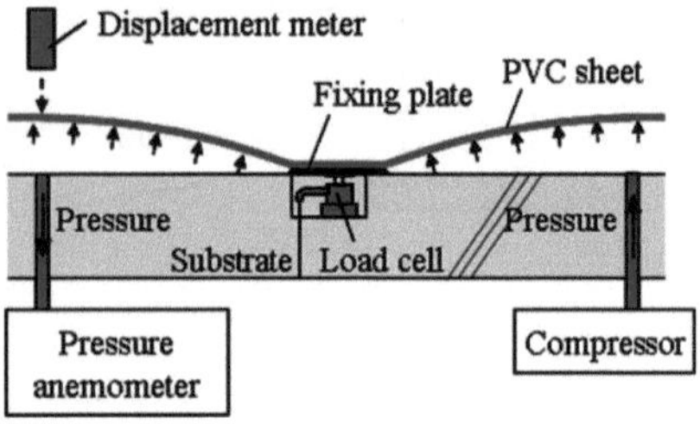

Fig. 6 Static pressure test method

Fig. 7 Billowing condition of waterproofing membrane during the static pressure test

The force induced in the fastener at point F in the pressure test is shown in Fig. 8. A horizontal line shows the force induced in the fastener calculated from the area of the waterproofing membrane that shares the load of one fastener. The load measured by the load cell and the equivalent load calculated by the above equation have a linear relationship and the transmission rate from the pressure to the load cell was 94 % at point F. Therefore, it was clarified that the transmission rate of the load from the pressure caused in the membrane to the fastener is high.

Relationships between the pressure and the billowing condition of the waterproofing membrane in the pressure test are shown in Fig. 9. The billowing heights of the waterproofing membrane at points H1 and H2 were approximately the same.

4 Test Results during TYHPOON No. 8 in 2009

4.1 Outline of Measurements

The following parameters were measured directly: wind direction, wind speed, wind pressure, the billowing of the waterproofing membrane, and axial-lateral force at fastener.

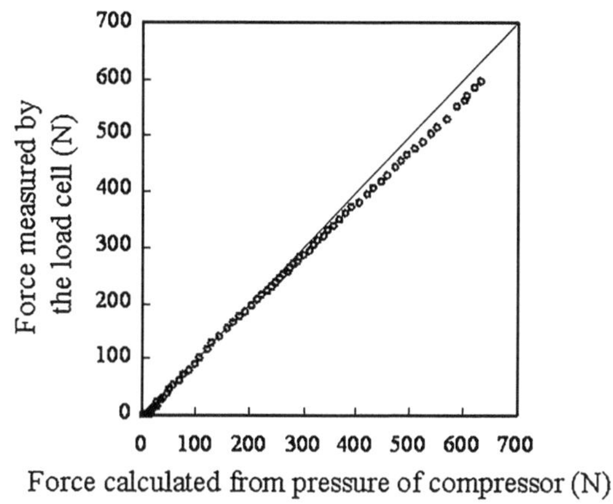

Fig. 8 Load transferred from wind pressure to load cell by pressure test of compressor

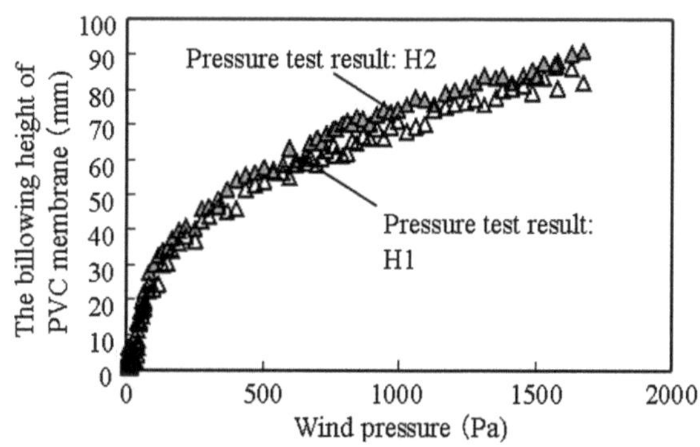

Fig. 9 Relationship between pressure and billowing height of PVC membrane

The equipment utilized for taking these measurements is as follows. We set up an ultrasound wind direction and wind speed meter, FT702 (measurement range: 0.01–60 m/s), 25 m away from the test building in the NW direction, 3 m above the ground (the same height as the test building).

The 25 wind pressure measurement points were established on the surface of the roof waterproofing sheet, denoted by P1 to P25, as shown in Fig. 5, and the method used to measure wind pressure is outlined in Fig. 10. The external pressure on the roof surface is conducted to a pressure anemometer, via an external pressure measurement pipe (positioned at the fixing plate) and a PVC tube of internal diameter 8 mm, thereby enabling the difference between this external pressure and reference wind pressure to be determined. As a waterproofing measure, we positioned the ends of the pipes 25 mm above the roof surface. In order to isolate the external pressure from the effects of rainwater, we used a Y-shaped wind pressure branching tube (so that rain water could be collected in a water tank).

We located the measurement points for billowing at the midpoints between locations where the waterproofing membrane was fixed to the roof, where billowing is expected to be greatest. These eight points, H1–H8, are shown in Fig. 5.

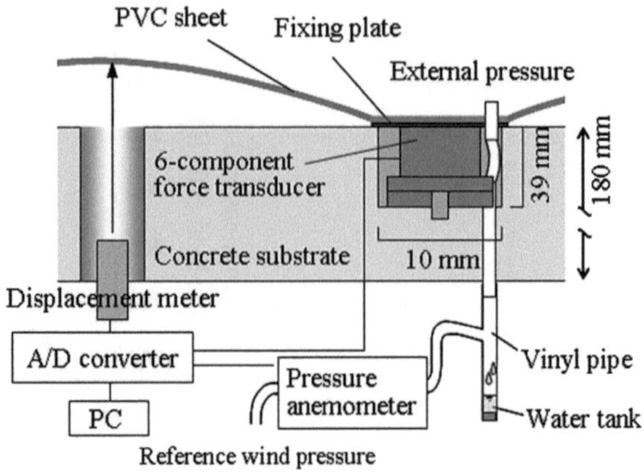

Fig. 10 The cross section of measurement method for the field test during Typhoon

To measure the membrane billowing, we drilled 40-mm holes in the concrete substrate, as shown in Fig. 10, and then measured the displacement relative to the bottom through these holes, using an ultrasound displacement meter.

The 6-component force transducer was set up to measure the axial force and lateral force at a fastener around waterproofing membrane. This 6-component force transducer enables simultaneous measurement of three forces (Fx, Fy, Fz) in three axial directions orthogonal to the transducer and three moments (Mx, My, Mz) around the axes.

Measurement data was recorded continuously, by observers who proceeded to the test locality as the typhoon approached Miyako Island, for the entire period during which the Miyako Island area was classified as being within the "strong wind zone." The data for each measurement point was sampled at 25 Hz for wind pressure and at 100 Hz for sheet billowing and strain, with all data saved to the hard disk of a PC. In addition, video footage of the roof behavior during exposure to strong wind was also recorded.

4.2 Wind Characteristics of TYHPOON No.8 in 2009

Between August 6 and 7, 2009, Typhoon No. 8, which was accompanied by heavy wind and rain, came as close as southwest of Miyako Island. The wind direction changed from NE to SE. The highest mean wind speed over 10 min was 18.8 m/s. Also, the wind direction and wind speed values measured at the test building corresponded closely with the figures from the meteorological observatory.

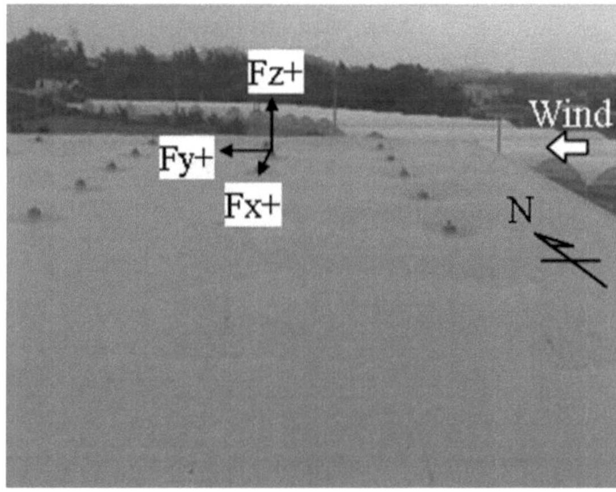

Fig. 11 The direction of Fx, Fy, Fz at 6 component transducer and the billowing condition of the PVC sheet during TYHPOON No.8 in 2009

4.3 Observation Results

Wind direction was from the east. Fig. 11 shows the billowing condition of the PVC sheet exposed during the 10-minute mean wind speed 15 m/s, the direction of Fx, Fy, Fz at the 6-component transducer, and the position of the PVC sheet billowing height. As a result of observation of the PVC sheet during the typhoon, we found that waterproofing membranes were sucked up by wind pressure and that the billowing was high on the windward side but relatively low on the leeward side. Moreover, in response to the strong wind, the PVC sheet appeared to exhibit a combination of two behaviors: billowing condition during constant billowing and an intermittent flapping in the direction of the wind. Therefore, the billowing height of the waterproofing membrane around the fasteners in the pressure test and the wind typhoon test was different.

4.4 Measurement Results and Discussion

Figure 12 shows the relationship between wind pressure and the billowing height of the PVC sheet. The billowing height of the PVC sheet was highest at the H4 point and was 40 mm at 15 m/s-mean wind speed. The billowing of the PVC sheet at windward was higher than that of leeward, and its value was 15 mm at H6 point. Then, the standard deviation means the fluttering condition of the PVC sheet, and the position of H4-H5-H6 was 5 mm to 7 mm.

Figure 13 shows the comparison between the lateral force (Fy) and the axial force (Fz). The lateral force and the axial force were linearly proportional to wind pressure and they increased with increased wind pressure. The Fz at the wind speed of 14.6 m/s was 40 N; by contrast, at Fy, the lateral force was 30 N, namely,

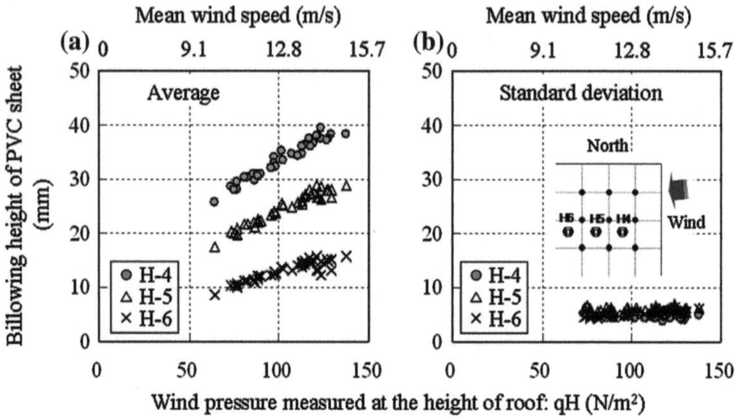

Fig. 12 Billowing height of PVC sheet during TYHPOON No.8 in 2009 (2009.8.7)

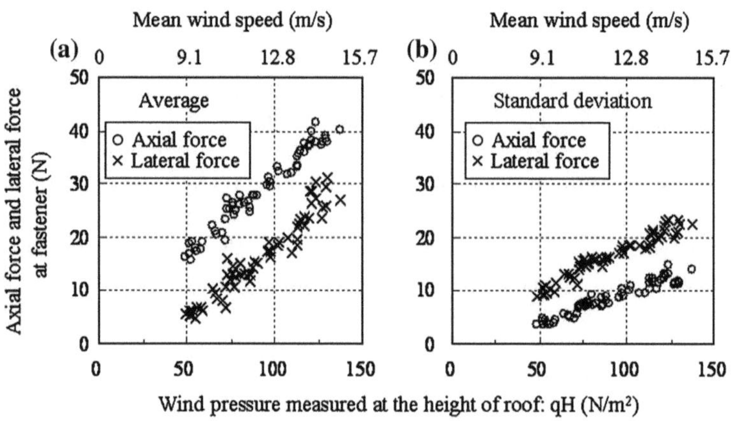

Fig. 13 Axial force and lateral force at fastener during TYHPOON No.8 in 2009 (2009.8.7)

about 75 % of the Fz, while the axial force was larger than the lateral force. However, the standard deviation of the lateral force (Fy) was higher than that of the axial force (Fz). Therefore, the fasteners were shaken back and forth or to the right and left by the strong winds. In case of the high wind pressure, it is necessary to consider not only the axial tension but also the lateral force for the design of the fastener.

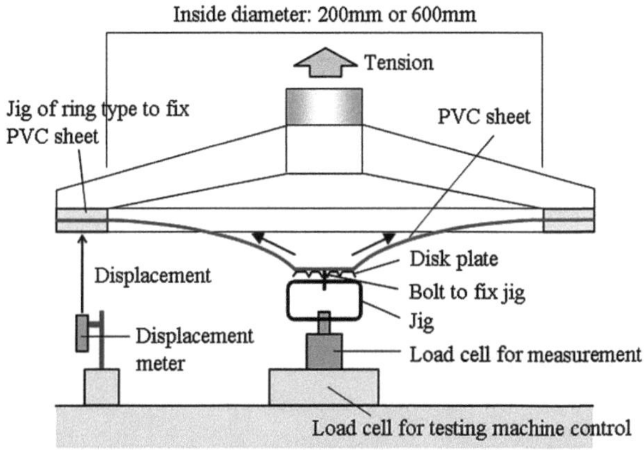

Fig. 14 Testing and measurement method by jig of ring type

5 Test Method with Circle Shaped Specimens

5.1 Method of Applying Repetitive Loads on Fastener of Membrane

Membranes are lifted by negative pressure caused by wind loads and the forces are transmitted to the fasteners as well as to the substrate. Our aim was to evaluate the actual fracture and the wind resistance of mechanically fastened waterproof systems. We assumed that the real conditions, applying negative pressure on and around membrane fastener, can be replaced by the mechanical test method. In the present experiment the simplified test method, where the actual wind suction was substituted by a jig attached to the membrane at its periphery as shown in Fig. 14, is proposed.

5.2 Circle Shaped Specimens

The circular specimen shown in Fig. 14 represents an easy tool for the evaluation of wind resistance of membrane fasteners. The edge of the membrane is sandwiched between the two steel circle shaped plates connected by bolts. By lifting the jig, repetitive loading is applied to the fastener.

Table 1 Testing condition by jig of ring type

Parameters	Testing contents
Water proofing type	-High strength type of PVC sheet -Low strength type of PVC sheet
Inside diameter of ring	-200 mm -600 mm
Loading type	-Static tensile test -Dynamic test(Uniform and Non-uniform load)

Fig. 15 Tensile test using jig of ring type

5.3 Tensile Test using Circle Shaped Specimens

5.3.1 Tensile Test Outline

The test overview is shown in Table 1. The two types of PVC membrane (high and low strength reinforcement) were put to the test. To verify the fastener distance as well as to confirm whether it is possible to make more compact specimens we prepared the specimens with dimensions of 600 and 200 mm.

An Instron type tensile and compression testing machine with the capacity of 500 kN was used. Tensile speed was set to 100 mm/min. The condition of circle shaped specimen during tensile test is shown in Fig. 15.

The test where only few types of mechanically fastened waterproofing system failures mentioned in Fig. 1 could occur was implemented; namely membrane tear, adhesive between the fixing plate and membrane failure, fixing plate failure.

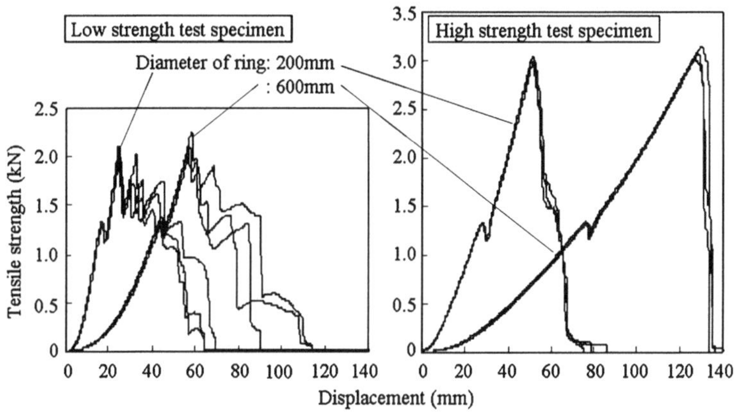

Fig. 16 Tensile test results using jig of ring type

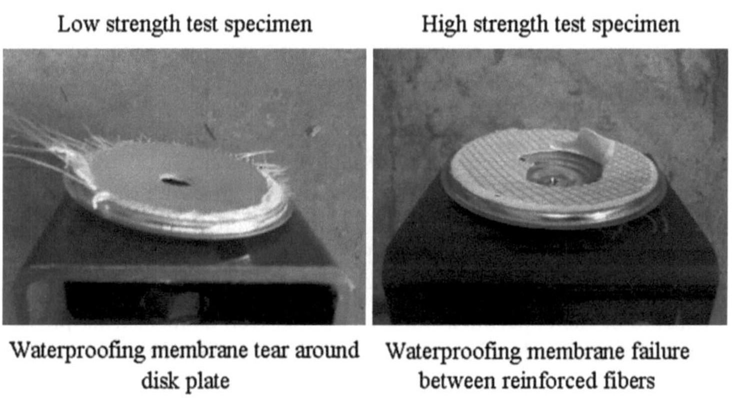

Fig. 17 Failure patterns of waterproofing membrane system in tensile test

5.3.2 Tensile Test Results

The relation between load and displacement during static tensile test is shown in Fig. 16. In specimens with both types of membrane reinforcement, before reaching maximum load, central part of fixing plate deformed and strength then slightly decreased. For low strength PVC sheets, average maximum load of 2103 and 2166 N was measured for the specimens with diameter 200 and 600 mm respectively. In both cases tear of the membrane occurred, as shown in the left side of the Fig. 17. High strength PVC sheets reached the average maximum load of 3021 and 3077 N for the specimens with diameter 200 and 600 mm respectively. In this case failure mode differed; membrane was torn in the position of the reinforcement fibers as shown on the right side of the Fig. 17. In this experiment, we could not confirm the effect of the fasteners distance of 200 and 600 mm.

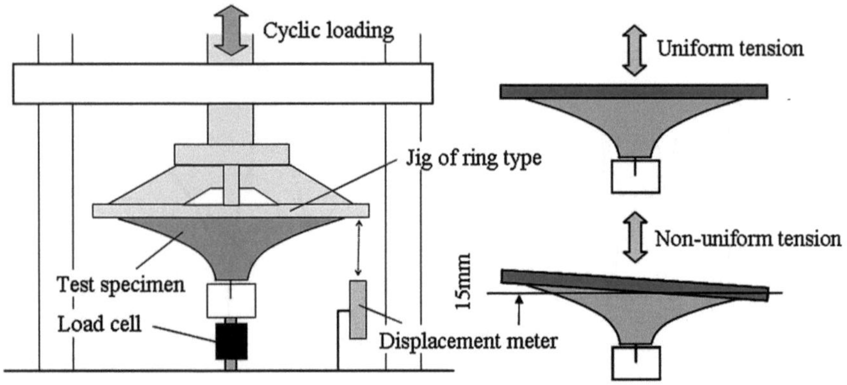

Figs. 18-19 Fig. 18: Cyclic loading test method and Fig. 19: cyclic loading pattern

5.4 Cyclic Fatigue Test

5.4.1 Outline of Fatigue Test

The same types of the PVC membranes and the specimens as during the static tensile test were tested. Hydraulic fatigue testing machine with maximum capacity of 100 kN was utilized. Circle shaped specimens were attached to the machine, and through the hole in the middle of the fixing plate they were attached to the testing jig as shown in Fig. 18. The jig was attached to the load cell (capacity 10 kN) capable of the experiment load control as well as measuring pressure and tensile loads. Repetitive loadings were applied to the fastener, and an ultrasonic displacement meter was installed to the bottom surface of the ring to measure the membrane deformation.

5.4.2 Test Conditions

Two types of the fatigue test were performed. Considering uneven billowing of waterproofing membrane caused by wind we performed (a) uniform tension test and (b) non-uniform tension test, which was achieved by inclining the ring type jig. This is described in Fig. 19. Similarly to the static test, fatigue tests of simulating fastener distances of 200 and 600 mm were performed. Specimen, load application method as well as measuring method are shown in Fig. 20

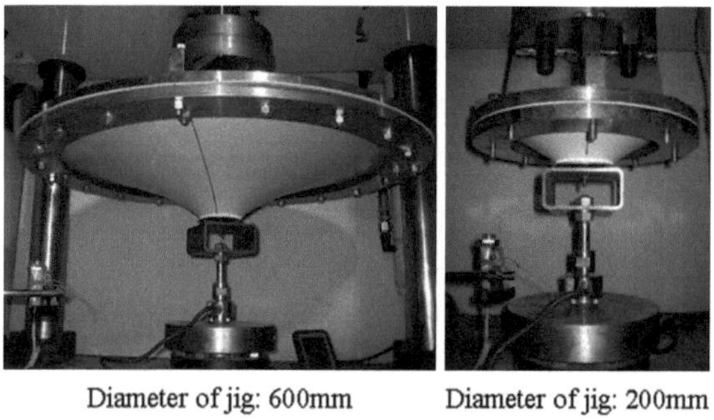

Fig. 20 Fatigue cyclic test condition

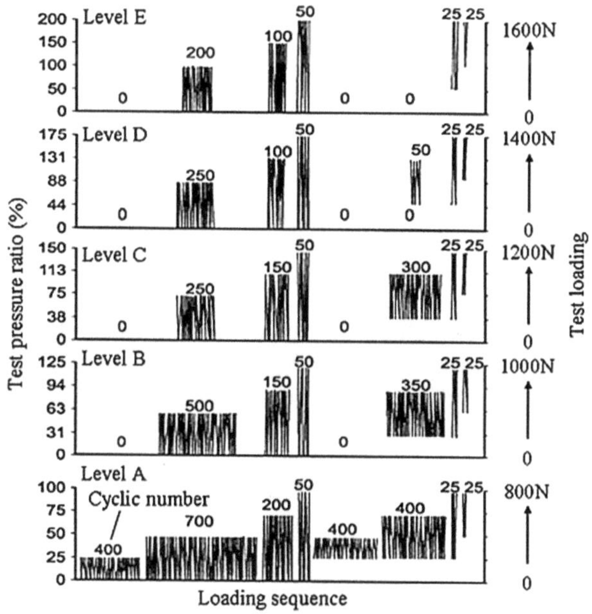

Fig. 21 SIGDERS dynamic wind uplift test protocol

5.4.3 Wind Resistance Evaluation Method

To perform the dynamic test, the SIGDERS procedure (Baskaran et al. 2009, 2008, 2007, 2005) was followed. The test method is described in Fig. 21. Wind resistance is evaluated by relation between design wind pressure rate and number of repetitions. Depending on wind pressure, test conditions can be set from the Level A to the highest level of the pressure, Level E (double loading conditions comparing to Level A). For instance, if the system passes the Level A of certain design wind pressure, the waterproofing system is defined as satisfactory for the 10 year

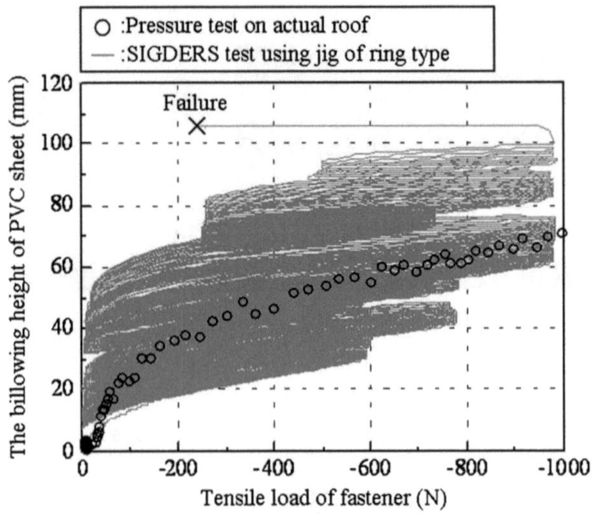

Fig. 22 Relationship between the pressure test and SIGDERS test (PVC sheet: Low strength, Diameter: 600 mm)

service life period for those wind pressure conditions. If the service life of waterproofing materials is evaluated by utilizing Level A protocol, the pressure is applied in four phases.

On the other hand, to evaluate the fatigue life in general, law of cumulative fatigue damage is often used [Society of Steel Construction of Japan (1993)]. To associate this idea to number of repetitions in the SIGDERS test procedure the following equations is used.

$$D = \frac{400}{N_{wf25}} + \frac{1100}{N_{wf50}} + \frac{600}{N_{wf75}} + \frac{100}{N_{wf100}} \tag{1}$$

where D is the degree of damage (D≧1: fracture occurs. D < 1: fracture does not occur), W_f is the wind pressure (Pa) and $N_{wf\ (N)}$ is the number of repetitions till fracture at W_f 25, 50, 75, 100 %.

In this experiment for the load control the SIGDERS test procedure was utilized. If the pressure W_f (Pa) set by SIGDERS test method is transformed to load per area unit (N/m^2), the load applied on one fastener can be calculated. Repetition period was set to once per 8 s.

5.4.4 Test Results of Experiment Utilizing SIGDERS Test Procedure

The relation between the tensile load and the membrane billowing height is shown in Fig. 22 for the SIGDERS test and in Fig. 9 for the static pressure test. Compared data in the figure for both experiments are the ones of fastener distance 600 mm. The

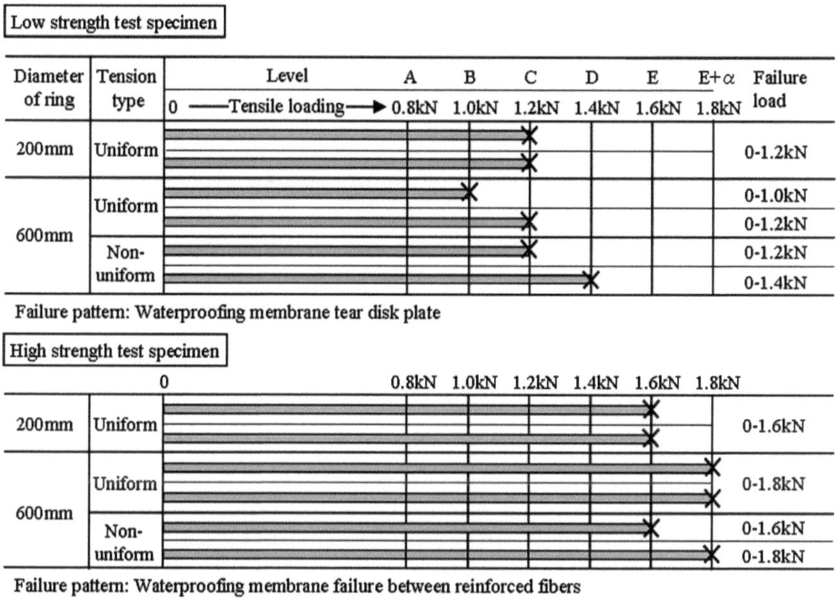

Fig. 23 Test results in SIGDERS test

SIGDERS test data show uniform tension fatigue history form Level A to Level B. As the tensile load increases, so does the billowing height of the membrane. As a criterion the tensile load level of 1 kN was set and the billowing heights of both specimens were compared. We obtained a good correlation between the results. In both cases the billowing height was 60 mm. During SIGDERS test as the repetitive loadings of 1 kN were applied to the membrane its billowing height increased gradually from 60 to 100 mm, clearly due to the fatigue effect.

Then, in Fig. 23 the results of the fatigue experiment utilizing SIGDERS procedure are presented. Fig. 24 shows the specimens' fracture modes. Identically to tensile test, low strength specimen's membrane was torn around the disk plate. In the case of high strength specimen failure mode differed; membrane was torn in the position of the reinforcement fibers. Regarding the fastener distance effect, the specimens with 600 mm diameter tend to reach higher fatigue resistance than the specimens of 200 mm.

In case of non-uniform loading specimens with inclined jig, in the early stage of the repetitive loadings the stresses were concentrated, but the membrane elongated initially and was able to withstand similar levels of loadings as applied to uniform specimens. Therefore, we could not confirm the difference.

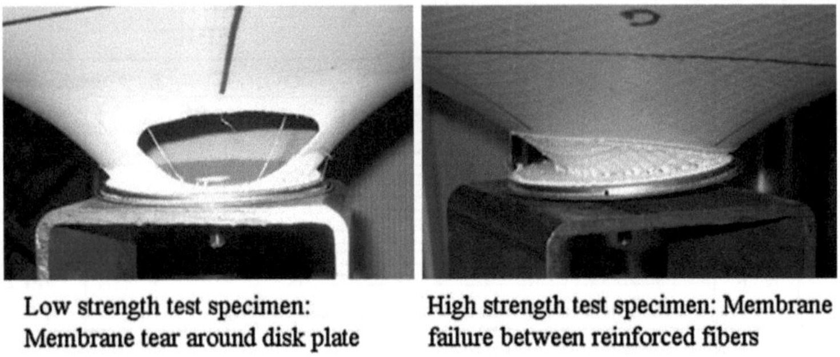

Fig. 24 Failure pattern

Fig. 25 Dynamic strength/ Static strength in each test condition

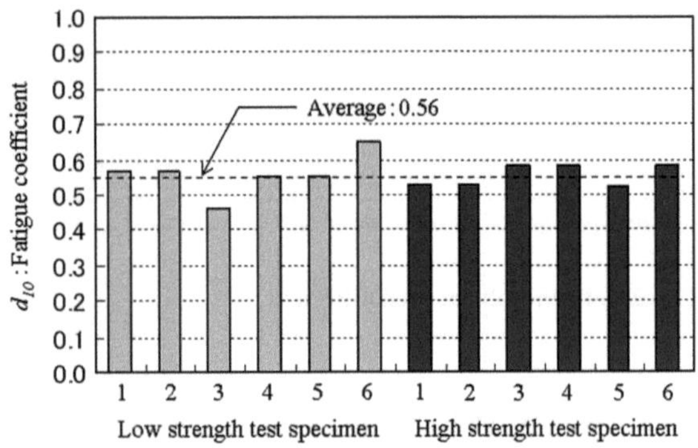

5.4.5 Differences between the Load Levels of Static and Dynamic Tests

During the service life of waterproofing membrane as it withstands repetitive loadings, the strength is gradually diminishing. In this research, from the test results utilizing SIGDERS procedure the fatigue degradation factor was calculated using following equation:

$$d_{10} = \frac{F_d}{F_s} \tag{2}$$

where d_{10} is the fatigue degradation factor, F_d is the load at which specimen failed in dynamic fatigue test (N) and F_s is the load at which specimen failed in static tensile test (N).

The fatigue degradation factor for each test is shown in Fig. 25. The factor d_{10} for the low strength specimens was in interval of 0.46 to 0.65. In case of the high strength specimens the values ranged from 0.52 to 0.58. The average value of both specimen types was 0.56. The effect of membrane reinforcement type appeared insignificant. The specimens' failure occurred at similar levels of loading.

6 Calculations for Number of Fasteners Considering Fatigue Effect

In conventional design of waterproofing systems' wind resistance, the fastener distance is determined considering the design wind loads above the roof. Also, as shown in Eq. 3, the empirical safety coefficient is considered. The values of safety coefficient vary widely from 2 to 5. We believe the verification of its values is important.

$$N = \frac{K \cdot W_f}{F} \tag{3}$$

where N is the number of fasteners required per unit area (pcs/m^2), W_f is the wind pressure (N/m^2), F is the design strength of fastener (N) and K is the Safety coefficient.

As a result of this research the Eq. 4 is presented. To follow SIGDERS evaluation procedure, for waterproofing membrane to pass the Level A, the service life of the membrane is defined and guaranteed for 10 years. Fig. 25 shows the comparison of the static and dynamic strengths. The safety coefficient covering the membrane 10 year service life can be replaced by the fatigue degradation factor. Hence,

$$N = \frac{W_f}{d_{10} \cdot F_s} \tag{4}$$

where d_{10} is the fatigue degradation factor of waterproofing membrane and F_s is the static tensile strength of circle shape specimen (N).

Relation of Eq. (4) to nomograph is shown in Fig. 26. For example, let us consider the high strength specimen with the static tensile strength $F_s = 3100$ N. On the left side of the figure using fatigue degradation factor ($d_{10} = 0.56$) slope line, the strength of the membrane after 10 years of exposure is 1736 N. In case of the wind pressure above the real roof, for example $W_f = 2800$ Pa, using Eq. (4) or the right side of Fig. 26 it results in the necessary number of fasteners higher than 1.6 pcs per area of one squared meter. Hence to fulfill the fatigue resistance demand, the number of fasteners per 1 m^2 becomes 2 pieces. Similarly, in case of low strength specimen with tensile strength of 2100 N, 2.4 fasteners per 1 m^2 is required, hence 3pieces/m^2 will be designed. From Eq. (4), not only period of use including the effect of fatigue and the fatigue degradation factor of the waterproofing membrane can be considered, but it also becomes possible to theoretically calculate the number of fasteners before the construction stage commences.

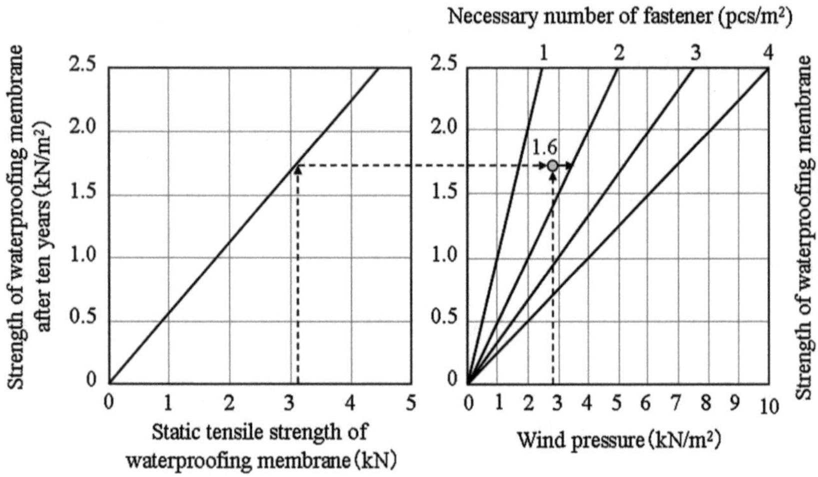

Fig. 26 Calculation for necessary number of fastener per unit area

7 Conclusions

The results of the tests conducted were as follows.

1. The transmission rate of load to the fasteners from pressure caused in the waterproofing membrane was 94 % in a static pressure test using a compressor, and therefore it was clarified that the transmission rate of load was high.
2. The billowing of the PVC sheet at windward was higher than that of leeward during typhoon. Therefore, the billowing height of the waterproofing membrane around the fasteners in the pressure test and the wind typhoon test was different.
3. The Fz at the wind speed of 14.6 m/s was 40 N; by contrast, at Fy, the lateral force was 30 N, namely, about 75 % of the Fz, while the axial force was larger than the lateral force. On the other hand, the standard deviation of the lateral force (Fy) was higher than that of the axial force (Fz). The higher the axial force, the higher the lateral force and the ratio of lateral force/axial force is increasing.
4. The comparison of the results of the static pressure test on test building with the results of the test using circle shaped specimens utilizing SIGDERS procedure shows very good correlation considering effects of the tensile strength and the membrane billowing height.
5. Test results of fatigue test using circle shaped specimens confirmed very small differences considering effects of the fastener distance and the membrane billowing height.

6. Results of the fatigue test show that, both low strength specimens and high strength specimens, failed at the level of 56 % of the static load of the static tensile test.
7. By performing tensile test and fatigue repetitive loading test applying the SIGDERS test procedure, we were able to present calculation method for the number of necessary fasteners per unit area defining the fatigue degradation factor of waterproofing membrane.

Acknowledgments The authors wish to thank Mr. Chea Daravuth and Mr. Yousuke Furusawa for their help with this project. This work was conducted as part of research activities of a working group for Assessment of Wind Endurance of Mechanically Anchored Waterproofing Membranes of the Waterproofing Membrane System of the Architectural Institute of Japan with the financial support of the following committee members: the Synthetic Polymeric Roofing Sheet Manufacturers Association (KRK), the Asphalt Waterproofing Manufacturers Association (ARK), the Torch-Applied Roofing Manufacturers Association (TRK), and fastener manufacturers. This work was supported by KAKENHI (20360249, Grant-in-Aid for Scientific Research (B)) in Japan, and Basic Science Research Program through the National Research Foundation of Korea (NRF) funded by the Ministry of Education, Science and Technology (2010-0011927). Some researchers were supported by Brain Korea 2nd (BK21) funded by the Korean Government. We are grateful to all these parties for their support.

References

Baskaran, B.A., Ko, S.K.P., Molleti, S.: A novel approach to estimate the wind uplift resistance of roofing systems. Build. Environ. **44**(4), pp. 723–735 (2009)

Baskaran, B.A., Ko, S., Molleti, S.: RCI continues its support for SIGDERS. phase V research and development activities. RC Items, pp. 4–5 (2008)

Baskaran, B.A., Paroli, R.M., Kalinger, P.: Advancements and changes in the North American commercial roofing industry. In: International Conference on Building Envelope Systems and Technology 2007, pp. 275–286, (2007)

Baskaran, B. A., Ko. S.K.P.: A guide for the wind design of mechanically attached flexible membrane roofs. Construction Innovation. Vol. 10, No. 4, NRC-IRC (2005)

Lei, W., Baskaran, B.A.: SIGDERS wind uplift resistance data on mechanically attached single ply roofing systems—effect of vapour/air barrier. Research Report IRC-RR-823 National Research Council, Canada (2001)

Baskaran, B.A., Chen, Y., Vilaipornsawai, U.: A new dynamic wind load cycle to evaluate flexible membrane roofs. ASTM J. Test. Eval. **27**(4), 249–265 (1999)

Uematsu, Y., Nakamura, S., Sera, M.: Wind pressure loading cycles for evaluating the wind resistance of mechanically attached waterproofing systems on flat roofs. In: 5th European and African Conference on Wind Engineering, pp. 375–378 (2009)

Furuichi, K., Uematsu, Y, Nakamura, S., Sera, M.: Evaluation of dynamic wind loads for mechanically-attached waterproofing membrane systems. Proceedings of 19th National Symposium on Wind Engineering, pp. 471–476 (2005)

Bartko, M., Miyauchi, H., Tanaka, K.: Fatigue resistance of mechanical fastened waterproofing systems to ALC substrate by repetitive loadings. J. Technol Des. AIJ **24**, 21–46 (2006)

Bartko, M., Miyauchi, H., Tanaka, K.: Development of wind resistance test method of mechanically anchored waterproofing system with ALC substrate. J. Struct. Constr. Eng. AIJ **598**, 7–12 (2005)

Ichikawa, H., Bartko, M., Kato, N., Miyauchi, H., Sasaki, T., Tanaka, K.: Behavior of mechanically anchored waterproofing membrane exposed to high wind speed of wind tunnel. J. Struct. Constr. Eng. AIJ **593**, 17–24 (2004)
Ichikawa, H., Katou, N., Miyauchi, H., Tanaka, K.: Wind tunnel tests of various mechanically anchored waterproofing membrane. J. Struct. Constr. Eng. AIJ **615**, 47–52 (2007)
Society of Steel Construction of Japan, Recommendations of Fatiuge Design for Steel Structures. p. 343, Gihodo (1993)

MIX
Papier aus verantwortungsvollen Quellen
Paper from responsible sources
FSC® C105338

If you have any concerns about our products,
you can contact us on
ProductSafety@springernature.com

In case Publisher is established outside the EU,
the EU authorized representative is:
Springer Nature Customer Service Center GmbH
Europaplatz 3, 69115 Heidelberg, Germany

Printed by Libri Plureos GmbH
in Hamburg, Germany